MEMOIRS
of the
American Mathematical Society

ISSN 0065-9266 (print)
ISSN 1947-6221 (online)

Number 1409

On Pseudoconformal Blow-Up Solutions to the Self-Dual Chern-Simons-Schrödinger Equation: Existence, Uniqueness, and Instability

Kihyun Kim
Soonsik Kwon

April 2023 • Volume 284 • Number 1409 (fifth of 6 numbers)

AMERICAN MATHEMATICAL SOCIETY

Library of Congress Cataloging-in-Publication Data

Cataloging-in-Publication Data has been applied for by the AMS.
See http://www.loc.gov/publish/cip/.
DOI: https://doi.org/10.1090/memo/1409

Memoirs of the American Mathematical Society

This journal is devoted entirely to research in pure and applied mathematics.

Subscription information. Beginning in 2023, Memoirs will be published monthly through 2026. Memoirs is also accessible from www.ams.org/journals. The 2023 subscription begins with volume 281 and consists of twelve mailings, each containing one or more numbers. Individual subscription prices for 2023 are as follows. For electronic only: US$984. For paper delivery: US$1,129. Add US$22 for delivery within the United States; US$85 for surface delivery outside the United States. Upon request, subscribers to paper delivery of this journal are also entitled to receive electronic delivery. For information on institutional pricing, please visit `https://www.ams.org/publications/journals/subscriberinfo`. Subscription renewals are subject to late fees. See `www.ams.org/journal-faq` for more journal subscription information. Each number may be ordered separately; please specify number when ordering an individual number.

Back number information. For back issues see `www.ams.org/backvols`.

Subscriptions and orders should be addressed to the American Mathematical Society, P. O. Box 845904, Boston, MA 02284-5904 USA. *All orders must be accompanied by payment.* Other correspondence should be addressed to 201 Charles Street, Providence, RI 02904-2213 USA.

Memoirs of the American Mathematical Society (ISSN 0065-9266 (print); 1947-6221 (online)) is published bimonthly (each volume consisting usually of more than one number) by the American Mathematical Society at 201 Charles Street, Providence, RI 02904-2213 USA. Periodicals postage paid at Providence, RI. Postmaster: Send address changes to Memoirs, American Mathematical Society, 201 Charles Street, Providence, RI 02904-2213 USA.

This publication is indexed in *Mathematical Reviews®*, *Zentralblatt MATH*, *Science Citation Index®*, *Science Citation Index™-Expanded*, *ISI Alerting Services℠*, *SciSearch®*, *Research Alert®*, *CompuMath Citation Index®*, *Current Contents®/Physical, Chemical & Earth Sciences*.
This publication is archived in *Portico* and *CLOCKSS*.
Printed in the United States of America.

∞ The paper used in this book is acid-free and falls within the guidelines established to ensure permanence and durability.
Visit the AMS home page at `https://www.ams.org/`

10 9 8 7 6 5 4 3 2 1   28 27 26 25 24 23 23

Contents

Abstract

We consider the self-dual Chern-Simons-Schrödinger equation (CSS), also known as a gauged nonlinear Schrödinger equation (NLS). CSS is L^2-critical, admits solitons, and has the pseudoconformal symmetry. These features are similar to the L^2-critical NLS. In this work, we consider pseudoconformal blow-up solutions under m-equivariance, $m \geq 1$. Our result is threefold. Firstly, we construct a pseudoconformal blow-up solution u with given asymptotic profile z^*:

$$\Big[u(t,r) - \frac{1}{|t|}Q\Big(\frac{r}{|t|}\Big)e^{-i\frac{r^2}{4|t|}}\Big]e^{im\theta} \to z^* \qquad \text{in } H^1$$

as $t \to 0^-$, where $Q(r)e^{im\theta}$ is a static solution. Secondly, we show that such blow-up solutions are unique in a suitable class. Lastly, yet most importantly, we exhibit an instability mechanism of u. We construct a continuous family of solutions $u^{(\eta)}$, $0 \leq \eta \ll 1$, such that $u^{(0)} = u$ and for $\eta > 0$, $u^{(\eta)}$ is a global scattering solution. Moreover, we exhibit a rotational instability as $\eta \to 0^+$: $u^{(\eta)}$ takes an abrupt spatial rotation by the angle

$$\Big(\frac{m+1}{m}\Big)\pi$$

on the time interval $|t| \lesssim \eta$.

We are inspired by works in the L^2-critical NLS. In the seminal work of Bourgain and Wang (1997), they constructed such pseudoconformal blow-up solutions. Merle, Raphaël, and Szeftel (2013) showed an instability of Bourgain-Wang solutions. Although CSS shares many features with NLS, there are essential differences and obstacles over NLS. Firstly, the soliton profile to CSS shows a slow polynomial decay $r^{-(m+2)}$. This causes many technical issues for small m. Secondly, due to the nonlocal nonlinearities, there are strong long-range interactions even between

---

Received by the editor July 22, 2019, and, in revised form, July 22, 2019.

Article electronically published on March 21, 2023.

DOI: `https://doi.org/10.1090/memo/1409`

2020 *Mathematics Subject Classification.* Primary 35B44, 35Q55.

Key words and phrases. Chern-Simons-Schrödinger equation, blow-up, pseudoconformal, self-duality, rotational instability.

The first author is affiliated with the Department of Mathematical Sciences, Korea Advanced Institute of Science and Technology, 291 Daehak-ro, Yuseong-gu, Daejeon 34141, Korea. Email: khyun1215@kaist.ac.kr.

The second author is affiliated with the Department of Mathematical Sciences, Korea Advanced Institute of Science and Technology, 291 Daehak-ro, Yuseong-gu, Daejeon 34141, Korea. Email: soonsikk@kaist.edu.

The authors were partially supported by Samsung Science & Technology Foundation BA1701-01 and National Research Foundation of Korea NRF-2019R1A5A1028324.

functions in far different scales. This leads to a nontrivial correction of our blow-up ansatz. Lastly, the instability mechanism of CSS is completely different from that of NLS. Here, the phase rotation is the main source of the instability. On the other hand, the self-dual structure of CSS is our sponsor to overcome these obstacles. We exploited the self-duality in many places such as the linearization, spectral properties, and construction of modified profiles.

CHAPTER 1

Introduction

The Chern-Simons-Schrödinger equations can be viewed as gauged nonlinear Schrödinger equations that generalize the one dimensional cubic nonlinear Schrödinger equation to the planar domain, with nonvanishing Chern-Simons gauge coupling. Interesting novel features are the *self-duality* and presence of vortex solitons. The goal of this work is to study a dynamical feature of this *gauged nonlinear Schrödinger equation* under self-duality. More specifically, we study pseudoconformal blow-up solutions. We construct a pseudoconformal blow-up solution for each given asymptotic profile (Theorem 1.1) and show that such a solution is unique (Theorem 1.2). Next, we exhibit a rotational instability of the blow-up solution (Theorem 1.3). Our work is inspired by works in the context of L^2-critical nonlinear Schrödinger equation, by Bourgain and Wang [**3**], and Merle, Raphaël, and Szeftel [**50**].

The Chern-Simons theory is a three dimensional topological gauge field theory whose action is an integral of Chern-Simons 3-form [**7**]. It has applications in condensed matter physics, knot theory, and low dimensional invariant theory.

As the Chern-Simons theory is formulated on three dimensional domain (space-time), it is applicable to describe the dynamics of particles confined to a spatial plane, so called *planar* physical phenomena, e.g. quantum Hall effect and high temperature superconductivity. This is a sharp contrast to the Yang-Mills or Maxwell theory that takes place on four dimensional space-time. In the 90's, relativistic and non relativistic Chern-Simons models were introduced to study vortex solutions in planar quantum electromagnetic dynamics. The Chern-Simons-Schrödinger equations [**17–20**] are nonrelativistic quantum models describing the dynamics of a large number of charged particles in the plane, that interact among themselves and the electromagnetic gauge field.

1.1. Covariant formulation

The Chern-Simons-Schrödinger model is a Lagrangian field theory on $\mathbb{R}^{1+2}$ associated to the action

$$\mathcal{S}[\phi, A] := \int_{\mathbb{R}^{1+2}} \Big(\frac{1}{2}\mathrm{Im}(\overline{\phi}\mathbf{D}_t\phi) + \frac{1}{2}|\mathbf{D}_x\phi|^2 - \frac{g}{4}|\phi|^4\Big) + \int_{\mathbb{R}^{1+2}} \frac{1}{2} A \wedge dA.$$

Here, $\phi : \mathbb{R}^{1+2} \to \mathbb{C}$ is a scalar field, $\mathbf{D}_\alpha := \partial_\alpha + iA_\alpha$ for $\alpha \in \{0,1,2\}$ is the covariant derivative, $A := A_0 dt + A_1 dx_1 + A_2 dx_2$ is the associated 1-form, $|\mathbf{D}_x\phi|^2 := |\mathbf{D}_1\phi|^2 + |\mathbf{D}_2\phi|^2$, and $g > 0$ is the strength of the nonlinearity.

In a more abstract fashion, one can view ϕ as a section of a $U(1)$ bundle over $\mathbb{R}^{1+2}$. As the topology of $\mathbb{R}^{1+2}$ is trivial, there is a global orthonormal frame, on which a metric connection $\mathbf{D}$ is viewed as $\mathbf{D} = d + iA$ for some real-valued 1-form A.

Computing the Euler-Lagrange equation, we obtain the *Chern-Simons-Schrödinger* equation

$$\begin{cases} \mathbf{D}_t\phi = i\mathbf{D}_j\mathbf{D}_j\phi + ig|\phi|^2\phi, \\ F_{01} = -\operatorname{Im}(\overline{\phi}\mathbf{D}_2\phi), \\ F_{02} = \operatorname{Im}(\overline{\phi}\mathbf{D}_1\phi), \\ F_{12} = -\frac{1}{2}|\phi|^2, \end{cases} \tag{1.1}$$

where $F_{jk} := \partial_j A_k - \partial_k A_j$ are components of the curvature 2-form $F := dA$. Repeated index j means that we sum over $j \in \{1, 2\}$. We note that

$$\mathbf{D}_\alpha \mathbf{D}_\beta - \mathbf{D}_\beta \mathbf{D}_\alpha = iF_{\alpha\beta}.$$

(1.1) is gauge-invariant. Indeed, if (ϕ, A) is a solution to (1.1) and $\chi : \mathbb{R}^{1+2} \to \mathbb{R}$ is any function, then

$$(e^{i\chi}\phi, A - d\chi) \tag{1.2}$$

is also a solution to (1.1).[1] When χ is a constant function on $\mathbb{R}^{1+2}$, we have the phase rotation symmetry.

In addition to gauge invariance, (1.1) enjoys many symmetries. It is invariant under space/time translations and spatial rotations. It also has time reversal symmetry[2] and Galilean invariance. Moreover, (1.1) enjoys scaling and pseudoconformal invariance. More precisely, assuming that (ϕ, A) is a solution to (1.1), so is $(\widetilde{\phi}, \widetilde{A})$:

- L^2*-scaling*: for any fixed $\lambda > 0$,

$$\begin{cases} \widetilde{\phi}(t, x) := \lambda\phi(\lambda^2 t, \lambda x), \\ \widetilde{A}_0(t, x) := \lambda^2 A_0(\lambda^2 t, \lambda x), \\ \widetilde{A}_j(t, x) := \lambda A_j(\lambda^2 t, \lambda x). \end{cases} \tag{1.3}$$

- *Pseudoconformal invariance*: (discrete form)

$$\begin{cases} \widetilde{\phi}(t, x) := \frac{1}{t} e^{i\frac{|x|^2}{4t}} \phi(-\frac{1}{t}, \frac{x}{t}), \\ \widetilde{A}_0(t, x) := \frac{1}{t^2} A_0(-\frac{1}{t}, \frac{x}{t}) - \frac{x_j}{t^2} A_j(-\frac{1}{t}, \frac{x}{t}), \\ \widetilde{A}_j(t, x) := \frac{1}{t} A_j(-\frac{1}{t}, \frac{x}{t}), \end{cases} \tag{1.4}$$

(or, in continuous form) for any fixed $a \in \mathbb{R}$,

$$\begin{cases} \widetilde{\phi}(t, x) := \frac{1}{1+at} e^{ia\frac{|x|^2}{4(1+at)}} \phi(\frac{t}{1+at}, \frac{x}{1+at}), \\ \widetilde{A}_0(t, x) := \frac{1}{(1+at)^2} A_0(\frac{t}{1+at}, \frac{x}{1+at}) - \frac{x_j}{(1+at)^2} A_j(\frac{t}{1+at}, \frac{x}{1+at}), \\ \widetilde{A}_j(t, x) := \frac{1}{1+at} A_j(\frac{t}{1+at}, \frac{x}{1+at}). \end{cases} \tag{1.5}$$

[1]Gauge invariance of (1.1) under (1.2) holds for an arbitrary χ. Under the additional assumption that χ is compactly supported in $\mathbb{R}^{1+2}$, the action $\mathcal{S}$ is gauge-invariant from the algebra $\int_{\mathbb{R}^{1+2}} d\chi \wedge dA = \int_{\mathbb{R}^{1+2}} d(\chi \wedge dA) = 0$. Although the Chern-Simons density $\frac{1}{2}(A \wedge dA)$ is *not* gauge-invariant itself, its integral *is* gauge-invariant. Note that the remaining densities are gauge-invariant.

[2]The time reversal symmetry is *not* simply given by conjugating the scalar field ϕ. In this paper, we use $\phi(t, x_1, x_2) \mapsto \overline{\phi}(-t, x_1, -x_2)$, $A_\alpha(t, x_1, x_2) \mapsto A_\alpha(-t, x_1, -x_2)$ for $\alpha \in \{0, 2\}$, and $A_1(t, x_1, x_2) \mapsto -A_1(-t, x_1, -x_2)$.

As the scaling invariance (1.3) preserves the L^2-norm of ϕ, we say that (1.1) is L^2-*critical*. The pseudoconformal invariance is a special feature of (1.1). In the context of nonlinear Schrödinger equations, pseudoconformal invariance holds only in the L^2-critical setting. It is natural to view (1.1) as a gauged L^2-critical nonlinear Schrödinger equation.

By Noether's principle, associated with phase and space-time translation invariance, the *charge* M, *momentum* J_j, and *energy* E are conserved:

$$\begin{aligned} M &:= \int_{\mathbb{R}^2} |\phi|^2, \\ J_j &:= \int_{\mathbb{R}^2} \mathrm{Im}(\overline{\phi}\mathbf{D}_j\phi), \\ E &:= \int_{\mathbb{R}^2} \Big(\frac{1}{2}|\mathbf{D}_x\phi|^2 - \frac{g}{4}|\phi|^4\Big). \end{aligned}$$

These are gauge-invariant quantities. There are other conservation laws, such as the angular momentum and normalized center of mass, associated with rotation and Galilean invariance. Finally, from the scaling and pseudoconformal symmetries, we have *virial identities*

$$\begin{cases} \partial_t\Big(\int_{\mathbb{R}^2} |x|^2|\phi|^2\Big) = 8\Phi, \\ \partial_t\Phi = 2E, \end{cases} \tag{1.6}$$

where

$$\Phi := \frac{1}{2}\int_{\mathbb{R}^2} \mathrm{Im}(\overline{\phi}\cdot x_j\mathbf{D}_j\phi).$$

One can alternatively write (1.6) as

$$8t^2E(e^{i\frac{|x|^2}{4t}}\phi(0)) = \int_{\mathbb{R}^2} |x|^2|\phi(t,x)|^2. \tag{1.7}$$

The aforementioned conservation laws can also be obtained from the *pseudo-stress-energy tensor*, which is defined by

$$\begin{aligned} T_{00} &:= \tfrac{1}{2}|\phi|^2, \\ T_{j0} = T_{0j} &:= \mathrm{Im}(\overline{\phi}\mathbf{D}_j\phi), \\ T_{jk} = T_{kj} &:= 2\mathrm{Re}(\overline{\mathbf{D}_j\phi}\mathbf{D}_k\phi) - \delta_{jk}(\Delta T_{00} + g|\phi|^2T_{00}). \end{aligned}$$

Here δ_{jk} is the Kronecker-delta and Δ is the Laplacian on $\mathbb{R}^2$. We have local conservation laws[3]

$$\begin{cases} \partial_t T_{00} + \partial_k T_{0k} = 0, \\ \partial_t T_{j0} + \partial_k T_{jk} = 0. \end{cases} \tag{1.8}$$

Taking space-time integrals of $T_{\alpha 0}$ with various weights, we obtain conservation laws.

[3]There is also a local conservation law for the energy. If we define the energy current by $e_0 := \frac{1}{2}|\mathbf{D}_x\phi|^2 - \frac{g}{4}|\phi|^4$ and $e_j := -\mathrm{Re}(\overline{\mathbf{D}_j\phi}\mathbf{D}_t\phi)$, then we have $\partial_t e_0 + \partial_j e_j = 0$.

Bogomol'nyi operator. Let $\widetilde{\mathbf{D}}_+$ be the *Bogomol'nyi operator* defined by

$$\widetilde{\mathbf{D}}_+ := \mathbf{D}_1 + i\mathbf{D}_2. \tag{1.9}$$

Its formal L^2-adjoint is given by $\widetilde{\mathbf{D}}_+^* = -\mathbf{D}_1 + i\mathbf{D}_2$. Observe using (1.1) that

$$\widetilde{\mathbf{D}}_+^*\widetilde{\mathbf{D}}_+ = -\mathbf{D}_j\mathbf{D}_j + F_{12} = -\mathbf{D}_j\mathbf{D}_j - \tfrac{1}{2}|\phi|^2. \tag{1.10}$$

Now the ϕ-evolution of (1.1) is rewritten as

$$i\mathbf{D}_t\phi - \widetilde{\mathbf{D}}_+^*\widetilde{\mathbf{D}}_+\phi + (g - \tfrac{1}{2})|\phi|^2\phi = 0. \tag{1.11}$$

Moreover, we can rewrite the energy functional as

$$E = \int_{\mathbb{R}^2}\Big(\frac{1}{2}|\widetilde{\mathbf{D}}_+\phi|^2 + \frac{1-g}{4}|\phi|^4\Big). \tag{1.12}$$

If $g < 1$, the energy is positive-definite. Thus it is natural to view (1.1) with $g < 1$ as a defocusing equation, for which the global well-posedness and scattering are expected. If $g > 1$, then (1.1) is viewed as a focusing equation that admits non-scattering (and even blow-up) solutions. We are interested in (1.1) with the critical coupling $g = 1$, which is referred to as the *self-dual* case. From Section 1.3, we restrict our discussions to the self-dual case.

Polar coordinates. Before imposing equivariant symmetry, it is convenient to formulate (1.1) in polar coordinates on $\mathbb{R}^2$. Let $r := |x|$, and define

$$\partial_r := \frac{1}{r}(x_1\partial_1 + x_2\partial_2) \quad\text{and}\quad \partial_\theta := -x_2\partial_1 + x_1\partial_2.$$

The corresponding covariant derivatives, connection, and curvature components become

$$\begin{aligned}
\mathbf{D}_r &= \tfrac{1}{r}(x_1\mathbf{D}_1 + x_2\mathbf{D}_2); & \mathbf{D}_\theta &= -x_2\mathbf{D}_1 + x_1\mathbf{D}_2;\\
A_r &= \tfrac{1}{r}(x_1A_1 + x_2A_2); & A_\theta &= -x_2A_1 + x_1A_2;\\
F_{0r} &= -\tfrac{1}{r}\mathrm{Im}(\overline{\phi}\mathbf{D}_\theta\phi); & F_{0\theta} &= r\mathrm{Im}(\overline{\phi}\mathbf{D}_r\phi);\\
F_{r\theta} &= rF_{12} = -\tfrac{r}{2}|\phi|^2.
\end{aligned}$$

The Bogomol'nyi operator has a simple expression in polar coordinates:

$$\widetilde{\mathbf{D}}_+ = e^{i\theta}[\mathbf{D}_r + \frac{i}{r}\mathbf{D}_\theta] \quad\text{and}\quad \widetilde{\mathbf{D}}_+^* = -e^{-i\theta}[\mathbf{D}_r - \frac{i}{r}\mathbf{D}_\theta].$$

In particular, we can rewrite the covariant Laplacian $\mathbf{D}_j\mathbf{D}_j$ as

$$\mathbf{D}_j\mathbf{D}_j = -\widetilde{\mathbf{D}}_+^*\widetilde{\mathbf{D}}_+ - \frac{1}{2}|\phi|^2 = \mathbf{D}_r\mathbf{D}_r + \frac{1}{r}\mathbf{D}_r + \frac{1}{r^2}\mathbf{D}_\theta\mathbf{D}_\theta.$$

Now the ϕ-evolution of (1.1) reads

$$i\mathbf{D}_t\phi + \Big(\mathbf{D}_r\mathbf{D}_r + \frac{1}{r}\mathbf{D}_r + \frac{1}{r^2}\mathbf{D}_\theta\mathbf{D}_\theta\Big)\phi + g|\phi|^2\phi = 0. \tag{1.13}$$

The energy takes the form

$$E = \int_{\mathbb{R}^2}\Big(\frac{1}{2}|\mathbf{D}_r\phi|^2 + \frac{1}{2r^2}|\mathbf{D}_\theta\phi|^2 - \frac{g}{4}|\phi|^4\Big). \tag{1.14}$$

1.2. Coulomb gauge and equivariance reduction

Recall that (1.1) has gauge invariance (1.2). Physically, we cannot observe ϕ and A by themselves, we only observe gauge-invariant quantities. Thus the evolution of (ϕ, A) described by (1.1) should be understood modulo gauge equivalence. In order to discuss the Cauchy problem of (1.1), we need to choose one representative (ϕ, A) from its (gauge-)equivalence class. This is usually established by imposing a gauge condition (or, fixing a gauge). In general, the Cauchy problem depends on the choice of gauge.

The Cauchy problem of (1.1) has been studied under the Coulomb gauge and heat gauge. Under the Coulomb gauge, Bergé-de Bouard-Saut [**1**] obtained local well-posedness in H^2. By a regularization argument, they also obtained global existence in H^1 for H^1 data having small charge, without uniqueness. Huh [**15**] showed that (1.1) has a unique-in-time solution for H^1 data, without continuous dependence though. Recently, Lim [**35**] obtained H^1 local well-posedness with local-in-time weak Lipschitz dependence. Using the heat gauge, Liu-Smith-Tataru [**37**] established local well-posedness in H^{ϵ}, $\epsilon > 0$, for small H^{ϵ} data with strong Lipschitz dependence. Under the equivariant symmetry (with Coulomb gauge), L^2-critical local theory is easily obtained solely using the Strichartz estimates by Liu-Smith [**36**].

There are also works on global-in-time behaviors. Pseudoconformal symmetry (1.4) applied to the static solution (1.26) directly gives a finite time blow-up solutions in the self-dual case $g = 1$ [**14, 17**]. On the other hand, Bergé-de Bouard-Saut [**1**] gave sufficient conditions on initial data yielding finite-time blow-up, based on virial identities (1.6) and a convexity argument. Oh-Pusateri [**51**] obtained small data *linear* scattering for (1.1) under the Coulomb gauge, by observing a cubic null structure. Using concentration-compactness arguments, Liu-Smith [**36**] showed global well-posedness and scattering for (1.1) under equivariance, if $g < 1$ or if the data has charge less than that of minimal standing waves when $g \geq 1$.

In this work, we impose the *Coulomb gauge condition*:

$$\partial_1 A_1 + \partial_2 A_2 = 0. \tag{1.15}$$

Differentiating curvature constraints (1.1) and using (1.15), we obtain

$$\begin{aligned} A_0 &= \Delta^{-1}[\epsilon_{jk}\partial_j \mathrm{Im}(\overline{\phi}\mathbf{D}_k\phi)], \\ A_j &= \tfrac{1}{2}\epsilon_{jk}\Delta^{-1}\partial_k|\phi|^2, \end{aligned}$$

where ϵ_{jk} is the anti-symmetric tensor with $\epsilon_{12} = 1$. In particular, the spatial part of A is given by the *Biot-Savart law*

$$A_j(t,x) = \frac{1}{4\pi}\epsilon_{jk}\int_{\mathbb{R}^2}\frac{(x_k - y_k)}{|x-y|^2}|\phi(t,y)|^2 dy. \tag{1.16}$$

As recognized in [**17**], (1.1) under the Coulomb gauge has a *Hamiltonian structure.* Indeed, it is a Hamiltonian flow associated to the energy E with the symplectic form $\omega(\phi,\psi) := \mathrm{Im}\int_{\mathbb{R}^2}\overline{\phi}\psi$. It is helpful to keep this structure in mind, as our duality estimates, self-dual form of the linearized operator, and expansion of the energy functional will all be intuitively based on the Hamiltonian structure.

Equivariance under the Coulomb gauge. Under the Coulomb gauge condition (1.15), we impose an *equivariance ansatz* as

$$\phi(t,x)=e^{im\theta}u(t,r) \tag{1.17}$$

for $m\in\mathbb{Z}$, where $x=x_1+ix_2=re^{i\theta}$. We call m the *equivariance index*. We will often write u in place of ϕ when there is no confusion. For example, we denote (1.19) and (1.20) below by $A_\theta[u]$ and $A_0[u]$. Or, the energy $E[\phi,A]=E[u,A[u]]$ is simply denoted by $E[u]$.

Under the equivariance ansatz (1.17), the connection components A_r, A_θ, and A_0 have simple expressions. Inserting (1.17) into the Biot-Savart law (1.16), we obtain

$$A_r=0. \tag{1.18}$$

Now A_θ can be obtained via integrating the curvature constraint from the origin:

$$A_\theta=-\frac{1}{2}\int_0^r |u|^2 r' dr'. \tag{1.19}$$

To get A_0, we integrate the curvature constraint from the infinity using the boundary condition $A_0(r)\to 0$ as $r\to\infty$:

$$A_0=-\int_r^\infty (m+A_\theta)|u|^2\frac{dr'}{r'}. \tag{1.20}$$

The Bogomol'nyi operator takes the form

$$\widetilde{\mathbf{D}}_+\phi=e^{i\theta}[\mathbf{D}_r+\frac{i}{r}\mathbf{D}_\theta]\phi=[(\partial_r-\frac{m+A_\theta}{r})u]e^{i(m+1)\theta}.$$

Taking the radial part, we write

$$\mathbf{D}_+u:=\partial_r u-\frac{m+A_\theta}{r}u. \tag{1.21}$$

Note that its formal L^2-adjoint is given by $\mathbf{D}_+^*u=-\partial_r u-\frac{m+1+A_\theta}{r}u$.

We sum up our discussions so far. Let A_r, A_θ, and A_0 be given by (1.18), (1.19), and (1.20), respectively. From (1.13), the ϕ-evolution is given by

$$i\partial_t\phi-A_0\phi+(\partial_{rr}+\frac{1}{r}\partial_r)\phi-\Big(\frac{m+A_\theta}{r}\Big)^2\phi+g|\phi|^2\phi=0.$$

Separating linear and nonlinear parts, we reorganize the above as

$$i\partial_t\phi+\Delta\phi=-g|\phi|^2\phi+\frac{2mA_\theta}{r^2}\phi+\frac{A_\theta^2}{r^2}\phi+A_0\phi. \tag{1.22}$$

The u-evolution is given by

$$i\partial_t u+\Delta_m u=-g|u|^2u+\frac{2mA_\theta}{r^2}u+\frac{A_\theta^2}{r^2}u+A_0u,$$

where Δ_m is the Laplacian adapted to m-equivariant functions

$$\Delta_m:=\partial_{rr}+\frac{1}{r}\partial_r-\frac{m^2}{r^2}. \tag{1.23}$$

From (1.12) and (1.14), the energy takes either the forms

$$E = \begin{cases} \int \Big(\frac{1}{2}|\partial_r u|^2 + \frac{1}{2}\Big(\frac{m+A_\theta}{r}\Big)^2|u|^2 - \frac{g}{4}|u|^4\Big), & \text{or} \\ \int \Big(\frac{1}{2}|\mathbf{D}_+ u|^2 + \frac{1-g}{4}|u|^4\Big). \end{cases}$$

Here, $\int$ denotes the integral $2\pi\int_0^\infty rdr$, as noted in (2.1).

1.3. Static solution Q for the self-dual (CSS)

From now on, we restrict to the self-dual case $g = 1$.

Substituting $g = 1$ into (1.12), the energy functional reads

$$E = \int_{\mathbb{R}^2} \frac{1}{2}|\widetilde{\mathbf{D}}_+\phi|^2,$$

provided $F_{12} = -\frac{1}{2}|\phi|^2$. In particular, the energy is always nonnegative. Here we explore the connection between zero-energy solutions and static solutions. It turns out that static solutions have zero energy and satisfies a first-order equation, and any zero-energy solutions are gauge equivalent to static solutions.

Let $\phi : \mathbb{R}^2 \to \mathbb{C}$ and $A_x = A_1 dx_1 + A_2 dx_2$ satisfy $F_{12} = -\frac{1}{2}|\phi|^2$. Then (ϕ, A_x) has zero energy if and only if $\widetilde{\mathbf{D}}_+\phi = 0$. We call the *Bogomol'nyi* (or, *self-dual*) *equation* as

$$\begin{cases} \widetilde{\mathbf{D}}_+\phi = 0, \\ F_{12} = -\frac{1}{2}|\phi|^2. \end{cases} \tag{1.24}$$

The self-duality can be seen from the identity [**17**]

$$\mathbf{D}_j\phi = -i\epsilon_{jk}\mathbf{D}_k\phi.$$

If we extend (ϕ, A) on $\mathbb{R}^{1+2}$ by requiring

$$\partial_t\phi = 0, \quad A_0 = \tfrac{1}{2}|\phi|^2, \quad \text{and} \quad \partial_t A = 0,$$

then the solution (ϕ, A) solves (1.1). As (ϕ, A) does not depend on time, such a solution is called *static*.

Consider now a solution (ϕ, A) to (1.1) having zero energy. It should satisfy the Bogomol'nyi equation (1.24) for each time. If we make a gauge transform such that $A_0 = \frac{1}{2}|\phi|^2$, then by (1.10) ϕ becomes static:

$$\partial_t\phi = -iA_0\phi + i\mathbf{D}_j\mathbf{D}_j\phi + i|\phi|^2\phi = -i\widetilde{\mathbf{D}}_+^*\widetilde{\mathbf{D}}_+\phi = 0.$$

Thus zero energy solutions are gauge equivalent to static solutions.

In fact, the converse direction holds. More precisely, if (ϕ, A) is a static solution to (1.1), then it satisfies (1.24), $A_0 = \frac{1}{2}|\phi|^2$, and has zero energy. To see this, (for a rigorous proof, see Huh-Seok [**16**]) we take the inner product to the static equation

$$A_0\phi - \mathbf{D}_j\mathbf{D}_j\phi - |\phi|^2\phi = 0 \tag{1.25}$$

with the covariant L^2-scaling vector field $(1 + x_k\mathbf{D}_k)\phi$ to conclude that (ϕ, A) has zero energy.[4] In particular, the Bogomol'nyi equation $\widetilde{\mathbf{D}}_+\phi = 0$ is satisfied. Applying (1.10), we have $A_0 = \frac{1}{2}|\phi|^2$.

[4]One should take care of the boundary terms arising from integration by parts. See [**16**].

We now impose the Coulomb gauge (1.15) and equivariance ansatz (1.17). We also concentrate on the *physically relevant case* $m \geq 0$ [**11**] from now on.

One can solve (1.24) following Jackiw-Pi [**17**]. Indeed, using (1.24), we have $\partial_r|\phi|^2 = \frac{2m+2A_\theta}{r}|\phi|^2$. If $|\phi(r_0)| = 0$ for some $r_0 \in (0,\infty)$, then $|\phi|(r) = 0$ for all $r \in (0,\infty)$ by the Gronwall inequality. Henceforth, we assume $|\phi| > 0$. Differentiating the curvature constraints, one can see that the charge density $|\phi|^2$ solves the Liouville equation

$$\Delta \log |\phi|^2 = -|\phi|^2.$$

General forms of solutions to the Liouville equation are known. There are explicit m-equivariant static solutions

$$\begin{cases} \phi^{(m)}(t,x) = \sqrt{8}(m+1)\dfrac{|x|^m}{1+|x|^{2(m+1)}}e^{im\theta}, \\ A_j^{(m)}(t,x) = 2(m+1)\dfrac{\epsilon_{jk}x_k|x|^{2m}}{1+|x|^{2(m+1)}}, \\ A_0^{(m)}(t,x) = 4\Big(\dfrac{(m+1)|x|^m}{1+|x|^{2(m+1)}}\Big)^2. \end{cases} \tag{1.26}$$

These solutions are unique up to the symmetries of (CSS); see [**5, 6, 8**]. Taking the radial part and suppressing the equivariance index m, we define[5]

$$Q(r) := \sqrt{8}(m+1)\frac{r^m}{1+r^{2(m+1)}}. \tag{1.27}$$

We have

$$\begin{aligned} M[Q] &= 8\pi(m+1), \\ E[Q] &= 0, \\ A_\theta[Q](r) &= -2(m+1)\frac{r^{2(m+1)}}{1+r^{2(m+1)}}. \end{aligned}$$

1.4. Pseudoconformal blow-up solutions and main results

We restrict ourselves (1.1) under the Coulomb gauge condition (1.15), equivariance ansatz (1.17), and the self-dual case $g = 1$. Therefore, we arrive at our main equation

$$i\partial_t u + \Delta_m u = -|u|^2 u + \frac{2m}{r^2}A_\theta u + \frac{A_\theta^2}{r^2}u + A_0 u, \tag{CSS}$$

where Δ_m is defined in (1.23) and the connection components are given by (1.18), (1.19), and (1.20):

$$\begin{cases} A_r = 0, \\ A_\theta = -\frac{1}{2}\int_0^r |u|^2 r' dr', \\ A_0 = -\int_r^\infty (m+A_\theta)|u|^2 \frac{dr'}{r'}. \end{cases}$$

Or, we can also rewrite (CSS) as a radial form

$$i\partial_t u = (-\partial_{rr} - \frac{1}{r}\partial_r)u + \Big(\frac{m+A_\theta}{r}\Big)^2 u + A_0 u - |u|^2 u. \tag{1.28}$$

---

[5]As mentioned above, we abuse notations such that $Q(x) = Q(r)e^{im\theta}$ denotes the m-equivariant extension of $Q(r)$.

Later, we will write (CSS) in a self-dual form (see (3.6))

$$i\partial_t u = L_u^* \mathbf{D}_+^{(u)} u.$$

The energy functional has either of the forms

$$E[u] = \begin{cases} \dfrac{1}{2}\displaystyle\int |\mathbf{D}_+ u|^2, \\ \dfrac{1}{2}\displaystyle\int |\partial_r u|^2 - \dfrac{1}{4}\displaystyle\int |u|^4 + \dfrac{1}{2}\displaystyle\int \Big(\dfrac{m + A_\theta}{r}\Big)^2 |u|^2. \end{cases} \tag{1.29}$$

In this work, we are going to construct a family of blow-up solutions. Applying the pseudoconformal symmetry (1.4) to the static solution Q, one obtains an explicit finite-time blow-up solution to (CSS) [**14, 17**]

$$S(t,r) := \frac{1}{|t|} Q\Big(\frac{r}{|t|}\Big) e^{-i\frac{r^2}{4|t|}}, \qquad \forall t < 0, \tag{1.30}$$

where we only wrote the radial part of the solution, for the sake of simplicity. One can apply various symmetries of (CSS) to obtain other explicit blow-up solutions.

Another blow-up solutions to (1.1) were considered by Bergé-de Bouard-Saut [**1**]. There, they used the virial identities and a convexity argument of Glassey [**12**] to give sufficient conditions for solutions of (1.1) to blow up in finite time (e.g. $E < 0$). Unfortunately, such a convexity argument is not so useful in the self-dual case $g = 1$, due to $E \geq 0$.[6] Indeed, if $u(t)$ is a solution to (CSS) with the initial data u_0 and satisfies $\int |x|^2 |u(t,x)|^2 = 0$ for some time $t \neq 0$, then $E[e^{i\frac{|x|^2}{4t}} u_0] = 0$ in view of (1.7). By the earlier discussion, u is the pseudoconformal transform of a static solution. This argument works in the covariant setting (1.1) with $g = 1$.

Let us call the blow-up rate of (1.30) as the *pseudoconformal blow-up rate*. By a *pseudoconformal blow-up solution*, we mean a finite-time blow-up solution having the pseudoconformal blow-up rate.

In this work, we study pseudoconformal blow-up solutions with prescribed asymptotic profiles z^*. We show existence, uniqueness, and instability. To state our result, let z^* be an m-equivariant profile satisfying the hypothesis

$$\begin{aligned} &-(m+2)\text{-equivariant function } \widetilde{z}^* := e^{-i(2m+2)\theta} z^* \text{ satisfies} \\ &\|\widetilde{z}^*\|_{H^k_{-(m+2)}} ¡ \alpha^* \text{ for some } k = k(m) > m+3, \end{aligned} \tag{H}$$

where H^k_ℓ is the space of ℓ-equivariant $H^k(\mathbb{R}^2)$ functions (see Section 2.2).

THEOREM 1.1 (Construction of pseudoconformal blow-up solutions). *Let $m \geq 1$. Let z^* be an m-equivariant profile satisfying* (H). *If $\alpha^* > 0$ is sufficiently small, then there is an m-equivariant solution u to* (CSS) *on $(-\infty, 0)$ with the property*

$$\Big[u(t,r) - \frac{1}{|t|} Q\Big(\frac{r}{|t|}\Big) e^{-i\frac{r^2}{4|t|}}\Big] e^{im\theta} \to z^* \qquad \text{in } H^1_m \tag{1.31}$$

as $t \to 0^-$. Indeed, u satisfies the following decomposition estimates

$$\begin{cases} \|u(t,r) - \dfrac{1}{|t|} Q_{|t|}\Big(\dfrac{r}{|t|}\Big) e^{i\gamma_{\mathrm{cor}}(t)} - z(t,r)\|_{\dot H^1_m} \lesssim \alpha^* |t|^m \quad \text{and} \\ \|u(t,r) - \dfrac{1}{|t|} Q_{|t|}\Big(\dfrac{r}{|t|}\Big) e^{i\gamma_{\mathrm{cor}}(t)} - z(t,r)\|_{L^2_m} \lesssim \alpha^* |t|^{m+1}, \end{cases} \tag{1.32}$$

[6] S.-J. Oh observed this fact. We are indebted to him for including this fact in our manuscript.

where $z(t,r)$ *is a solution to* (zCSS) *with the initial data* $z(0,r) = z^*(r)$ *constructed in Section* 5.1 *and* $\gamma_{\mathrm{cor}}(t)$ *is defined in* (5.23).

The solution constructed in Theorem 1.1 is unique in the following sense.

THEOREM 1.2 (Conditional uniqueness of the pseudoconformal blow-up solutions). *Let* $m \geq 1$. *Let* z^* *be an* m*-equivariant profile satisfying* (H). *Assume two* H^1_m*-solutions* u_1 *and* u_2 *to* (CSS) *satisfy*

$$\|u_j(t,r) - \frac{1}{|t|}Q_{|t|}\Big(\frac{r}{|t|}\Big)e^{i\gamma_{\mathrm{cor}}(t)} - z(t,r)\|_{H^1_m} \leq c|t| \tag{1.33}$$

for all $j = 1, 2$ *and* t *near zero, for sufficiently small* $\alpha^* > 0$ *and* $c > 0$. *Then* $u_1 = u_2$.

Lastly, we show that the pseudoconformal blow-up solution is instable. This immediately follows by constructing an one-parameter family of solutions in the following sense.

THEOREM 1.3 (Instability of pseudoconformal blow-up solutions). *Let* $m \geq 1$. *Assume the hypothesis of Theorem* 1.1. *Let* u *be the solution constructed in Theorem* 1.1. *There exist* $\eta^* > 0$ *and one-parameter family of* H^1_m*-solutions* $\{u^{(\eta)}\}_{\eta \in [0,\eta^*]}$ *to* (CSS) *with the following properties.*

- $u^{(0)} = u$,
- *For* $\eta > 0$, $u^{(\eta)}$ *scatters both forward and backward in time,*
- *The map* $\eta \in [0,\eta^*] \mapsto u^{(\eta)}$ *is continuous in the* $C_{(-\infty,0),\mathrm{loc}}H^{1-}$ *topology.*
- *The family* $\{u^{(\eta)}\}_{\eta\in[0,\eta^*]}$ *exhibits the rotational instability near time* 0*: we can write*

$$u^{(\eta)}(t,x) = \frac{e^{im(\theta + \gamma^{(\eta)}(t))}}{\langle t \rangle} Q^{(\eta)}_{|t|}\Big(\frac{r}{\langle t \rangle}\Big) + O_{H^1_m}(\alpha^*),$$

where $\langle t \rangle := (|t|^2 + \eta^2)^{\frac{1}{2}}$, $Q^{(\eta)}$ *is some profile defined in Proposition* 4.4, *and* $\gamma^{(\eta)}$ *satisfies*

$$|\gamma^{(0)}(-\tau)| \lesssim \alpha^* \tau,$$

$$\limsup_{\eta \to 0+} \Big|\gamma^{(\eta)}(\tau) - \gamma^{(\eta)}(-\tau) - \Big(\frac{m+1}{m}\Big)\pi\Big| \lesssim \alpha^*\tau,$$

for all small $\tau > 0$.

Taking the pseudoconformal transform on the family $\{u^{(\eta)}\}_{\eta\in[0,\eta^*]}$, we get

COROLLARY 1.4. *Let* $m \geq 1$. *Let* $\{u^{(\eta)}\}_{\eta\in[0,\eta^*]}$ *be as in Theorem* 1.3. *Let* $\mathcal{C}$ *be the pseudoconformal transformation defined in* (1.4). *Then,*

- $\mathcal{C}u^{(0)} - Q$ *scatters forward in time.*
- $\mathcal{C}u^{(\eta)}$ *scatters forward in time.*

What can be seen at a glance is that $\mathcal{C}u^{(0)} - Q$ scatters to a linear solution of $i\partial_t + \Delta_{-m-2}$, but $\mathcal{C}u^{(\eta)}$ scatters to that of $i\partial_t + \Delta_m$. However, the sense of scattering in both cases turn out to be equivalent. See Remark 5.6.

To our best knowledge, Theorem 1.1 provides the first example of finite-time blow-up solutions other than $S(t)$ in the context of (CSS). As mentioned earlier, this work is inspired by the seminal work of Bourgain-Wang [**3**] in the context of the nonlinear Schrödinger equation, in which they constructed pseudoconformal blow-up solutions for given asymptotic profiles. As (CSS) is a L^2-critical Schrödinger-type equation, it shares many features with L^2-critical NLS. It is worthwhile to review the results in NLS and compare them with (CSS).

The mass-critical *nonlinear Schrödinger equation* (NLS) on two dimension is

$$i\partial_t \psi + \Delta \psi + |\psi|^2 \psi = 0, \tag{NLS}$$

where $\psi : I \times \mathbb{R}^2 \to \mathbb{C}$. (NLS) has no static solutions, but has only standing waves of the form

$$(t, x) \mapsto \lambda e^{i\lambda^2 t} R(\lambda x), \qquad \forall \lambda > 0,$$

where R is the ground state satisfying

$$\Delta R - R + R^3 = 0. \tag{1.34}$$

Applying the pseudoconformal symmetry (1.4), we have an explicit pseudoconformal blow-up solution for (NLS)

$$S_{\mathrm{NLS}}(t, x) := \frac{1}{|t|} R\Big(\frac{x}{|t|}\Big) e^{\frac{i}{|t|}} e^{-i\frac{|x|^2}{4|t|}}, \qquad \forall t < 0. \tag{1.35}$$

Bourgain-Wang [**3**] constructed a family of pseudoconformal blow-up solutions to (NLS):

THEOREM 1.5 (Bourgain-Wang solutions [**3**], stated informally). *Let $\zeta^* : \mathbb{R}^2 \to \mathbb{C}$ be a profile that degenerates at the origin at large order, i.e. $|\zeta^*(x)| \lesssim |x|^A$ for some large A, and lies in some weighted Sobolev space. Then, there exists a (conditionally unique) solution ψ_{BW} to* (NLS) *defined near time zero such that*

$$\psi_{\mathrm{BW}}(t) - S_{\mathrm{NLS}}(t) \to \zeta^*$$

as $t \to 0$.

Our Theorem 1.1 is analogous to Theorem 1.5.

Pseudoconformal blow-up solutions are believed to be non-generic. In the context of (NLS), there are two supplementary works in this direction. Krieger-Schlag [**28**] showed that for the 1D critical NLS, there is a *codimension 1* manifold of initial data in the measurable category, yielding a pseudoconformal-type blow-up. Merle-Raphaël-Szeftel [**50**] exhibited instability of Bourgain-Wang solutions [**3**] as follows.

THEOREM 1.6 (Instability of Bourgain-Wang solutions [**50**], stated informally). *Let ζ^* be as in Theorem* 1.5 *and small in some weighted Sobolev space; let ψ_{BW} be the associated Bourgain-Wang solution. Then, there exists a continuous family of solutions ψ_η to* (NLS) *for $\eta \in [-\delta, \delta]$ with $0 < \delta \ll 1$ such that*

- *$\psi_0 = \psi_{\mathrm{BW}}$ is the Bourgain-Wang solution,*
- *For $\eta > 0$, ψ_η is global in time and scatters both forward and backward in time.*
- *For $\eta < 0$, ψ_η scatters backward and blows up forward in finite time under the log-log law* (1.37).

They perturbed R to the instability direction ρ_{NLS} (given in Remark 3.5), say $R - \eta\rho_{\mathrm{NLS}}$, and derived dynamical laws of modulation parameters depending on the sign of η.

Although it is not present in Theorem 1.6, it is worthnoting that they considerably relaxed the degeneracy asssumption on ζ^* at the origin. They also improved the notion of uniqueness of ψ_{BW} that is necessary for Theorem 1.6. Unlike [**3**], they rely on a robust modulation analysis and avoid the explicit use of the pseudoconformal symmetry. However, the pseudoconformal symmetry is implicitly used in the sense that they made approximate profiles by conjugating the phase $e^{-ib\frac{|y|^2}{4}}$. Our work also uses modulation analysis as in [**50**].

Theorem 1.3 is analogous to Theorem 1.6. However, the instability mechanism of (CSS) turns out to be quite different from that of (NLS); the instability of (CSS) stems from the phase rotation. This is due to the difference of the spectral properties for each problem.

Comparison between (CSS) *and* (NLS).

We view (CSS) as a gauged (NLS). Thus it is instructive to compare fundamental facts and results with (NLS).

1. *Symmetries and conservation laws.* All the symmetries of (CSS) (and also (1.1)) are valid for (NLS), including the scaling, Galilean, and pseudoconformal symmetries. Conservation laws of (CSS) are valid for (NLS) as well, after replacing the covariant derivatives by the usual derivatives. Virial identities also hold.

2. *Linearized operators around Q or R.* As we have seen above, (CSS) admits a static solution Q, but (NLS) only admits a standing wave solution $e^{it}R(x)$, which is not static. The mass term $-R$ is present in (1.34), but not in the static equation (1.25). From this, R decays exponentially at spatial infinity, but Q only decays as $r^{-(m+2)}$.

The linearized dynamics of (NLS) near $e^{it}R$ and that of (CSS) near Q differ drastically. Denote by $\mathcal{L}_{\mathrm{NLS}}$ the linearized operator for (NLS)

$$\mathcal{L}_{\mathrm{NLS}}f = -\Delta f + f - 2R^2 f - R^2\overline{f} = L_+\mathrm{Re}(f) + iL_-\mathrm{Im}(f). \tag{1.36}$$

$\mathcal{L}_{\mathrm{NLS}}$ has continuous spectrum $[1,\infty)$ and two eigenvalues, 0 and a negative value. On the other hand, the linearized operator $\mathcal{L}_Q$ (see (3.7)-(3.8)) is non-negative definite with $\Lambda Q, iQ \in \ker \mathcal{L}_Q$, where Λ is the generator of L^2-scaling (2.2). More importantly, (CSS) has the *self-dual* structure. By (3.8), this appears at the linearized level as

$$\mathcal{L}_Q = L_Q^* L_Q,$$

where L_Q is a first-order nonlocal differential operator defined by (3.3). The self-duality plays a crucial role in several places. It helps us to analyze the spectral property and coercivity of $\mathcal{L}_Q$ under generic orthogonality conditions. It is also used in the construction of the modified profiles $Q^{(\eta)}$ in Chapter 4. The factorization of the linearized operator was crucially used in blow-up problems of other evolution equations; see for instance [**54, 58**] for wave maps and [**49**] for Schrödinger maps, and [**55, 56**] for the energy-critical heat flow.

Let us restrict $\mathcal{L}_{\mathrm{NLS}}$ on radial functions. It is known from [**62**] that

$$\mathcal{L}_{\mathrm{NLS}}\Lambda R = -2R.$$

On the other hand, the linearized operator $\mathcal{L}_Q$ (see (3.7) and (3.8)) satisfies

$$\mathcal{L}_Q\Lambda Q = 0.$$

This difference is intimately tied to the fact that $e^{it}R$ is not a static solution to (NLS), but Q is a static solution to (CSS). As a consequence, the generalized null spaces of $i\mathcal{L}_{\mathrm{NLS}}$ and $i\mathcal{L}_Q$ appear to be rather different. Compare Proposition 3.4 with Remark 3.5.

3. *Dynamics below the threshold mass.* R and Q have *threshold mass*[7] in the sense that nontrivial dynamical behaviors arise from that level of mass. Any solutions to (NLS) having mass less than that of R are global-in-time [**61**] and scatter [**9, 27**]. In the context of (CSS) (under equivariance), Liu-Smith [**36**] proved the analogous result, where Q plays the role of R. At the level of the threshold mass, we have pseudoconformal blow-up solutions $S(t)$ and $S_{\mathrm{NLS}}(t)$. There are partial results on the threshold dynamics of (NLS); see for instance [**26, 34, 42**]. There is no related results for (CSS).

4. *Stable blow-up law.* Finite-time blow-up solutions (with negative energy) have been studied extensively in the context of (NLS). The energy of (NLS) can now attain negative values. In a seminal series of papers by Merle-Raphaël [**43, 45–48, 53**], they obtained a sharp description of blow-up dynamics for H^1 solutions to (NLS) having slightly super-critical mass and negative energy; there exists a universal constant c_* such that $\psi(t)$ blows up in the *log-log law* (almost self-similar with the log-log correction)

$$\|\nabla\psi(t)\|_{L^2} \approx c_*\Big(\frac{\log|\log(T-t)|}{T-t}\Big)^{\frac{1}{2}} \tag{1.37}$$

as t approaches to the blow-up time T. In fact, there is a larger open set of data yielding the log-log law. (See also a former work by Perelman [**52**]) It is also known from their results that any finite-time blow-up solutions with slightly supercritical mass must have blow-up rate either the log-log law, or faster than or equal to the pseudoconformal rate. For (1.1) with $g > 1$, when energy is negative, it is expected that a similar log-log law holds; see Bergé-de Bouard-Saut [**2**].

Note that the blow-up solution S_{NLS} does not follow the log-log law. It has threshold mass and is apparently instable with respect to shrinking its mass slightly. The solution S to (CSS) also has threshold mass and is instable by the same reason.

Comments on Theorems 1.1-1.3.

1. *Assumption* (H). Among smooth functions, (e.g. in the space of m-equivariant Schwartz functions) assumption (H) is equivalent to saying that z^* is small with $(\partial_r)^m z^*(0) = 0$. Indeed, a smooth m-equivariant function z^* satisfies $(\partial_r)^\ell z^*(0) = 0$ for either $0 \le \ell < m$ or $\ell - m$ is odd. Thus the condition $(\partial_r)^m z^*(0) = 0$ can be considered as a codimension 1 condition in the space of asymptotic profiles.

Assumption (H) requires z^* to be degenerate at the origin. Indeed, any smooth m-equivariant function f exhibits degeneracy $|f(x)| \lesssim r^{|m|}$ (see Section 2.2) at the origin, so a $-(m+2)$-equivariant function with $m \ge 1$ shows the degeneracy at least r^3. Such a degeneracy is necessary to decouple the marginal interaction between S and z^* because $S(t)$ concentrates at the origin as $t \to 0$. In (NLS), Bourgain-Wang [**3**] and Merle-Raphaël-Szeftel [**50**] used the flatness of the asymptotic profile ζ^* to construct pseudoconformal blow-up solutions.

The choice $-(m+2)$ arises naturally from the *long-range interaction* in (CSS). As opposed to the (NLS) case, (CSS) has long range potential as can be easily seen

[7] In fact for (CSS), $M[u]$ is called charge. In this discussion, we also call $M[u]$ mass.

from the Biot-Savart law (1.16) or the term

$$\Big(\frac{m + A_\theta[u]}{r}\Big)^2 u$$

of (1.28). Regardless how fast u decays, the above potential exhibits r^{-2} tail in general. In our case, our blow-up ansatz is $u \approx S + z$ (recall S (1.30)) so there is an interaction term

$$\Big(\frac{m + A_\theta[S]}{r}\Big)^2 z.$$

As S is concentrated near the origin, we may approximate $A_\theta[S](r)$ by $A_\theta[S](+\infty) = -2(m+1)$. Now the choice $(m+2)$ of (H) arises from

$$(\partial_{rr} + \frac{1}{r}\partial_r) - \Big(\frac{m - 2(m+1)}{r}\Big)^2 = (\partial_{rr} + \frac{1}{r}\partial_r) - \Big(\frac{-(m+2)}{r}\Big)^2 = \Delta_{-m-2}.$$

This observation is crucial in our work and leads to a *modification* of z-evolution (zCSS). This will be discussed in more detail in Section 5.1.

2. *Assumption $m \geq 1$.* There are many places that the proof does not work if $m = 0$. We list a few here.

Note that the solution $S(t)$ belongs to H^1 if and only if $m \geq 1$. The assumption $m \geq 1$ is best possible to construct solutions satisfying (1.31) without truncating the blow-up profile. In this work, we will heavily rely on the explicit computations coming from the pseudoconformal phase, so we hope to avoid any truncations.

There are extra simplifications coming from $m \geq 1$. The equivariant Sobolev space $\dot{H}^1_m$ (see Section 2.2) has a nice embedding property $\|\frac{f}{r}\|_{L^2} + \|f\|_{L^\infty} \lesssim \|f\|_{\dot{H}^1_m}$, which fails if $m = 0$. Moreover, the first order operator L_Q (given in (3.3)) shows strong coercivity $\|L_Q f\|_{L^2} \gtrsim \|f\|_{\dot{H}^1_m}$ under suitable orthogonality conditions on f for $m \geq 1$. If $m = 0$, the coercivity becomes weaker; see Lemma 3.9.

Assumption $m \geq 1$ reduces the bulk of the paper and make our arguments easier to follow, both conceptually and technically. Of course, the case $m = 0$ remains as an interesting open problem.

3. *Interaction of $S(t)$ and z^*.* A sharp contrast to (NLS) is that (CSS) has the long-range interaction even though $S(t)$ and z^* live in completely different scales. In the setup of our blow-up ansatz, we have to detect and take the strong interaction into account. This leads to a modification of the z-evolution and a correction of the phase parameter $\gamma(t)$. This is one of the novelties of this work. Even in NLS class, if the nonlinearity is replaced by a nonlocal nonlinearity, there could be a long range interaction. In [**30**], the authors observed a long range interaction and constructed Bourgain-Wang type solutions for the 4D mass-critical Hartree equation.

However, it turns out that the interaction is not strong enough to affect the blow-up rate. In general, the regularity and decay of the asymptotic profile is related to the blow-up rate. It is an interesting question whether one can generate exotic blow-up rates by prescribing other types of z^*. There are many works on constructing exotic blow-up solutions, for example, [**10, 22, 29, 31, 32, 40**]. There are also works exhibiting strong interactions between bumps in different scales, for instance, construction of multi-bubbles; see [**21, 23–25, 41**].

4. *On uniqueness.* The time-decay condition (1.33) of Theorem 1.2 is tight with (1.32) of Theorem 1.1 when $m = 1$. (1.33) is best from our method. When m is large, another argument presented in [**50**] (see also [**57**]) works similarly for (CSS). However, when m is 1 or 2, the argument does not work due to a slow decay

of Q. We instead redo the construction proof on the difference $\epsilon = \epsilon_1 - \epsilon_2$ with a careful choice of modulation parameters.

5. *Rotational Instability.* The source of the instability comes from the phase rotation. This is a sharp contrast to that in (NLS). Mathematically, the difference comes from that of spectral properties of the linearized operators. A striking feature is that $u^{(0)} = u$ does not rotate at all, but $u^{(\eta)}$, $0 < \eta \ll 1$, shows an abrupt spatial rotation in a very short time interval $|t| \lesssim \eta$, by the angle

$$\Big(\frac{m+1}{m}\Big)\pi.$$

In the energy-critical Schrödinger map (1-equivariant), it is believed that there is a rotational instability of blow-up solutions. In [**49**], authors perform a slow-adiabatic ansatz both in scaling and rotation parameters to construct smooth blow-up solutions in the codimension 1 sense. Their proof sheds light on the rotational instability of the blow-up solutions. However, a sharp description on instability is not fully understood.

Note that we exhibited an one-sided instability ($\eta \geq 0$) of pseudoconformal blow-up solutions. The sign condition $\eta \geq 0$ is used crucially in two places of our proof: the construction of $Q^{(\eta)}$ and Lyapunov functional method.

1.5. Strategy of the proof

Our main theorems are to show the existence, uniqueness, and instability of pseudoconformal blow-up solutions. We will use modulation analysis as in [**50**]. Unlike [**3**], we do not explicitly use the pseudoconformal transform, but instead use the pseudoconformal phase $e^{-ib\frac{|y|^2}{4}}$.

With a prescribed asymptotic profile z^*, we set $t = 0$ as the blow-up time and construct u on $[t_0^*, 0)$. This is often called a *backward construction.* We will show the instability and existence at the same time by constructing a family of solutions $\{u^{(\eta)}\}_{\eta\in(0,\eta^*]}$ as in Theorem 1.3. Then, the blow-up solution $u^{(0)}$ is obtained by a limiting argument as $\eta \to 0^+$. The limiting argument requires two steps: obtaining uniform estimates on η and performing a soft compactness argument. This kind of argument is originally due to Merle [**44**] in the context of (NLS). Of course, there should be a separate argument for the uniqueness of the blow-up solution to conclude the instability (to guarantee continuity of the map $\eta \in [0, \eta^*] \mapsto u^{(\eta)}$). From now on, we focus on the construction of $u^{(\eta)}$ (Theorem 1.3).

1. *Setup for the modulation analysis.* For each fixed $\eta \in (0, \eta^*]$, we write our solution $u^{(\eta)}(t, x)$ as

$$u^{(\eta)}(t,x) = \frac{e^{i\gamma(t)}}{\lambda(t)}(Q^{(\eta)}_{b(t)} + \epsilon)(t, \frac{x}{\lambda(t)}) + z(t,x)$$

with the initial data

$$u^{(\eta)}(0,x) = \frac{1}{\eta}Q^{(\eta)}\Big(\frac{x}{\eta}\Big) + z^*(x).$$

Here, $Q^{(\eta)}$ is some modified profile of Q to be explained soon, $f_b(y) := f(y)e^{-ib\frac{|y|^2}{4}}$ for a function f, and $z(t, x)$ is some function with $z(0, x) = z^*(x)$. We want to construct $u^{(\eta)}$ such that

$$\lambda(t) \approx \langle t\rangle = (t^2 + \eta^2)^{\frac{1}{2}}, \quad b(t) \approx |t|, \quad \text{and} \quad \gamma(t) \approx (m+1)\tan^{-1}\Big(\frac{t}{\eta}\Big). \tag{1.38}$$

The motivation for the above choice of modulation parameters will be explained later. Note that the spatial angle $(\frac{m+1}{m})\pi = \frac{1}{m}(\gamma(0+) - \gamma(0-))$ comes from the identity $e^{i\gamma}e^{im\theta} = e^{im(\theta+\frac{\gamma}{m})}$. If we take the limit $\eta \to 0^+$ for each fixed $t < 0$, we get the pseudoconformal regime.[8]

We will use the rescaled variables (s, y) such that

$$\frac{ds}{dt} = \frac{1}{\lambda^2} \quad \text{and} \quad y = \frac{x}{\lambda}.$$

Note that $Q_b^{(\eta)}$ and ϵ are functions of (s, y), but u and z are functions of (t, x). We often switch between the original variables (t, x) and rescaled variables (s, y) via $\sharp$ and $\flat$ notations. For example,

$$\epsilon^\sharp(t, x) = \frac{e^{i\gamma(t)}}{\lambda(t)}\epsilon(s(t), \frac{x}{\lambda(t)}) \quad \text{and} \quad z^\flat(s, y) = e^{-i\gamma(s)}\lambda(s)z(t(s), \lambda(s)y).$$

For parameters λ, γ, and b, we abbreviated $\lambda(t(s)) = \lambda(s)$ and so on. In Section 2.5, we provide conversion formulae between the (t, x)-variables and (s, y)-variables.

2. *Linearization of* (CSS). Since we use a decomposition in the (s, y)-variables as

$$u^\flat = Q_b^{(\eta)} + \epsilon + z^\flat,$$

it is essential to investigate the linearized evolution around $w^\flat := Q_b^{(\eta)} + z^\flat$. As $z^\flat$ is assumed to be small, we may replace $w^\flat$ by $Q_b^{(\eta)}$. As b and η are small, $Q_b^{(\eta)}$ is a modified profile of Q. Thus we need to study the linearized operator around Q, say $\mathcal{L}_Q$. In view of *self-duality* ($g = 1$), it turns out that the linearized operator $\mathcal{L}_Q$ is factorized as

$$\mathcal{L}_Q = L_Q^* L_Q$$

for some nonlocal first-order differential operator L_Q. This is first observed by Lawrie-Oh-Shahshahani [**33**]. This factorization is crucial in analyzing the spectral properties of $\mathcal{L}_Q$, including coercivity. Moreover, we are able to invert $\mathcal{L}_Q$ under a suitable solvability condition, even if $\mathcal{L}_Q$ is a *nonlocal* second-order differential operator. As a result, we are able to compute the generalized nullspace of $i\mathcal{L}_Q$ with the relations

$$\begin{aligned} i\mathcal{L}_Q\rho &= iQ; & \qquad i\mathcal{L}_Q ir^2Q &= 4\Lambda Q; \\ i\mathcal{L}_Q iQ &= 0; & \qquad i\mathcal{L}_Q\Lambda Q &= 0, \end{aligned}$$

where Λ is the L^2-scaling vector field defined in (2.2), and ρ is some profile. It is instructive to compare this with (NLS), see Remark 3.5.

Coming back to the nonlinear level, self-duality appears in the energy functional as

$$E[u] = \frac{1}{2}\int |\mathbf{D}_+u|^2.$$

In view of the Hamiltonian structure, we can write (CSS) in a more compact form

$$i\partial_t u = L_u^*\mathbf{D}_+^{(u)}u,$$

where L_u is the linearized operator of $\mathbf{D}_+u$ and L_u^* is its dual. See (3.6). This compact form simplifies the analysis technically. Moreover, it represents that we implicitly use Hamiltonian structure throughout the analysis.

---

[8]In fact, since $\tan^{-1}(\frac{t}{\eta}) \to -\frac{\pi}{2}$ as $\eta \to 0^+$, we get a phase-rotated pseudoconformal blow-up solution by the angle $-(\frac{m+1}{2})\pi$.

3. *Profile* $Q^{(\eta)}$. Recall the expression of $S(t)$ (1.30)

$$S(t,x)=\frac{1}{|t|}Q_{|t|}\Big(\frac{x}{|t|}\Big).$$

Thus S is an exact solution to (CSS) with $\epsilon=z=0$ and $\eta=0$. Assuming $\epsilon=z=0$ for a moment, the equation of $u^\flat$ and the ansatz $u^\flat=Q_{b(s)}$ suggest

$$\frac{\lambda_s}{\lambda}+b=0,\quad b_s+b^2=0,\quad \gamma_s=0.$$

Note that $\lambda(t)=b(t)=|t|$ and $\gamma(t)=0$ indeed solve the above ODEs.

One of novelties of this work is finding a rotational instability mechanism by introducing the η-parameter to γ_s. Assume $\frac{\lambda_s}{\lambda}+b=0$ and $Q_b^{(\eta)\sharp}$ solves (CSS). Then, we have

$$L_{Q^{(\eta)}}^{*}\mathbf{D}_{+}^{(Q^{(\eta)})}Q^{(\eta)}+\gamma_s Q^{(\eta)}-(b_s+b^2)\tfrac{|y|^2}{4}Q^{(\eta)}=0.$$

It turns out that we can impose

$$\gamma_s=\eta\theta_\eta=\eta\Big(\frac{1}{2}\int|Q^{(\eta)}|^2rdr-(m+1)\Big)\quad\text{and}\quad b_s+b^2+\eta^2=0.$$

See Section 4.2 for the motivation. This leads to

$$L_{Q^{(\eta)}}^{*}\mathbf{D}_{+}^{(Q^{(\eta)})}Q^{(\eta)}+\eta\theta_\eta Q^{(\eta)}+\eta^2\tfrac{|y|^2}{4}Q^{(\eta)}=0.$$

This is a *nonlocal* second-order equation. Due to the presence of $\eta\theta_\eta Q^{(\eta)}$, one expects $Q^{(\eta)}$ decays exponentially. But Q itself shows a polynomial decay, and hence it is difficult to approximate $Q^{(\eta)}(r)$ by an η-expansion of $Q(r)$ for r large. Such an η-expansion is not successful especially when m is small. We thus have to search for a nonlinear ansatz $Q^{(\eta)}$. Here, the self-duality plays a crucial role. We essentially use the self-duality to reduce this to a first-order equation what we call the *modified Bogomol'nyi equation*

$$\mathbf{D}_{+}^{(Q^{(\eta)})}P^{(\eta)}=0,\quad Q^{(\eta)}=e^{-\eta\frac{r^2}{4}}P^{(\eta)}.$$

The modified profile $Q^{(\eta)}$ is found by solving this nonlocal first-order equation. The choice (1.38) is obtained by integrating the ODEs

$$\frac{\lambda_s}{\lambda}+b=0,\quad \gamma_s=\eta\theta_\eta,\quad b_s+b^2+\eta^2=0.$$

4. *Derivation of corrections.* As mentioned earlier, although $Q_b^{(\eta)\sharp}$ and z live on far different scales, there are strong interactions between $Q_b^{(\eta)\sharp}$ and z from the nonlocal nonlinearities. Before performing a perturbative analysis, we need to capture and incorporate them into the blow-up ansatz. Recall (1.28). It turns out that there are two types of corrections: one is from $Q_b^{(\eta)\sharp}$ to z and the other is from z to $Q_b^{(\eta)\sharp}$. The former is absorbed by modifying the z-evolution and the latter is absorbed by correcting the phase parameter γ_s by the amount $\theta_{z^\flat\to Q_b^{(\eta)}}$. It is also interesting that the modified z-evolution becomes $-(m+2)$-equivariant (CSS). This is a sharp contrast to (NLS) since it is from the nonlocal nonlinearities. See more details in Section 5.1.

5. *Setup for bootstrapping.* Combining all the above information, we arrive at the ϵ-equation

$$\begin{aligned} &i\partial_s \epsilon - \mathcal{L}_{w^\flat}\epsilon + ib\Lambda\epsilon - \eta\theta_\eta\epsilon \\ &= i\Big(\frac{\lambda_s}{\lambda} + b\Big)\Lambda(Q_b^{(\eta)} + \epsilon) + (\widetilde{\gamma}_s - \eta\theta_\eta)Q_b^{(\eta)} + (\gamma_s - \eta\theta_\eta)\epsilon \\ &\quad - (b_s + b^2 + \eta^2)\tfrac{|y|^2}{4}Q_b^{(\eta)} + \widetilde{R}_{Q_b^{(\eta)},z^\flat} + V_{Q_b^{(\eta)}-Q_b}z^\flat + R_{u^\flat - w^\flat}. \end{aligned}$$

Here, $\widetilde{\gamma}_s = \gamma_s + \theta_{z^\flat \to Q_b^{(\eta)}}$ and $\widetilde{R}_{Q_b^{(\eta)},z^\flat}$ represents the interaction between $Q_b^{(\eta)}$ and $z^\flat$ after deleting the aforementioned strong interactions. The term $V_{Q_b^{(\eta)}-Q_b}z^\flat$ is originated from the difference of $Q^{(\eta)}$ and Q. Now $R_{u^\flat - w^\flat}$ collects the quadratic and higher terms in ϵ. These are all treated perturbatively. Among these, $\widetilde{R}_{Q_b^{(\eta)},z^\flat}$ is estimated by exploiting the decoupling in scales of $Q_b^{(\eta)}$ and $z^\flat$.

So far, we have not specified the choice of (λ, γ, b). To fix them, we will require two orthogonality conditions and one dynamical law governing λ and b. We impose two orthogonality conditions to guarantee the coercivity

$$(\epsilon, \mathcal{L}_Q \epsilon)_r \gtrsim \|\epsilon\|_{\dot{H}^1_m}^2.$$

Because of the factorization $\mathcal{L}_Q = L_Q^* L_Q$, a generic choice of orthogonality conditions suffices. For the last one, we choose to impose a *dynamical law* of λ and b:

$$2\Big(\frac{\lambda_s}{\lambda} + b\Big)b - (b_s + b^2 + \eta^2) = 0. \tag{1.39}$$

We are motivated to make this choice to cancel out $|y|^2 Q_b^{(\eta)}$ terms of the ϵ-equation since $|y|^2 Q_b^{(\eta)}$ has insufficient decay (uniformly in η) for small m. More precisely, in the ϵ-equation we have

$$\begin{aligned} &i\Big(\frac{\lambda_s}{\lambda} + b\Big)\Lambda Q_b^{(\eta)} - (b_s + b^2 + \eta^2)\tfrac{|y|^2}{4}Q_b^{(\eta)} \\ &= i\Big(\frac{\lambda_s}{\lambda} + b\Big)[\Lambda Q^{(\eta)}]_b + \Big[2\Big(\frac{\lambda_s}{\lambda} + b\Big)b - (b_s + b^2 + \eta^2)\Big]\tfrac{|y|^2}{4}Q_b^{(\eta)}. \end{aligned}$$

Note that the first term with $[\Lambda Q^{(\eta)}]_b$ has a sufficient decay.

Having fixed the dynamical laws of λ, γ, b, and ϵ, we setup the bootstrap procedure for λ, b, γ, $\|\epsilon\|_{\dot{H}^1_m}$, and $\|\epsilon\|_{L^2}$. By standard modulation estimates, we can control the variation of λ, b, and γ by ϵ. We also need a separate argument to control $\|\epsilon\|_{L^2}$ by $\|\epsilon\|_{\dot{H}^1_m}$. The usual energy inequality does not suffice, we use the Strichartz estimates to control L^2 norm of ϵ. On the way, we obtain estimates on $\|\epsilon\|_{L^2}$ by integrating in time. For this purpose, we introduce a weighted time maximal function $\mathcal{T}_{\dot{H}^1_m}^{(s,\eta)}[\epsilon]$.

6. *Lyapunov/virial functional method.* To close the bootstrap argument, we need to control $\dot{H}^1_m$ norm of ϵ. From the choice of our initial data, we have $\epsilon(0,x) = 0$. We will propagate smallness from $t = 0$ to $t = t_0^*$ via the Lyapunov method. Such an energy method was first used by Martel [**38**] in a backward construction. We will construct a functional $\mathcal{I}$ such that $\mathcal{I}$ controls the $\dot{H}^1_m$ norm of ϵ, and $\partial_t \mathcal{I}$ is almost nonnegative (as we evolve backward in time). To exploit the coercivity,

a natural candidate for $\mathcal{I}$ would be the quadratic (and higher) parts of the energy functional

$$E_{w^\flat}^{(\mathrm{qd})}[\epsilon] = E[w^\flat + \epsilon] - E[w^\flat] - (\frac{\delta E}{\delta u}\Big|_{u=w^\flat}, \epsilon)_r.$$

Solely using $E_{w^\flat}^{(\mathrm{qd})}[\epsilon]$, we cannot have sufficient monotonicity. This is due to the scaling and mass terms $ib\Lambda\epsilon - \eta\theta_\eta\epsilon$, i.e. ϵ essentially evolves under

$$i\partial_s\epsilon - \mathcal{L}_{w^\flat}\epsilon + ib\Lambda\epsilon - \eta\theta_\eta\epsilon \approx 0.$$

To incorporate $ib\Lambda\epsilon$, we add $b\Phi_A[\epsilon]$ to $E_{w^\flat}^{(\mathrm{qd})}[\epsilon]$, where Φ_A is a deformation of Φ in terms of truncation weight function ϕ_A (see (2.20)). $b\Phi_A$ is called a *virial correction,* which was first introduced by Raphaël-Szeftel [**57**]. To incorporate $-\eta\theta_\eta\epsilon$, we further add $\frac{\eta\theta_\eta}{2}M[\epsilon]$. As a result, we will consider

$$\mathcal{I}_A := \lambda^{-2}(E_{w^\flat}^{(\mathrm{qd})}[\epsilon] + b\Phi_A[\epsilon] + \tfrac{\eta\theta_\eta}{2}M[\epsilon]).$$

With this functional $\mathcal{I}_A$, we are able to prove that $\lambda^2\mathcal{I}_A - \frac{\eta\theta_\eta}{2}M[\epsilon] \sim \|\epsilon\|_{\dot{H}_m^1}^2$ from the coercivity of $\mathcal{L}_Q$. For $\lambda^2\partial_s\mathcal{I}_A$, we roughly have

$$\lambda^2\partial_s\mathcal{I}_A \approx 2b(E_Q^{(A),(\mathrm{qd})}[\epsilon] + \tfrac{\eta\theta_\eta}{2}M[\epsilon]),$$

where $E^{(A)}$ is a deformation of E in terms of truncation weight function ϕ_A. Note that b has positive sign. We have

$$\begin{aligned} E_Q^{(A),(\mathrm{qd})}[\epsilon] \approx{}& \frac{1}{2}\int \mathbf{1}_{r\leq A}|L_Q\epsilon|^2 \\ &+ \frac{1}{2}\int \mathbf{1}_{r>A}\phi_A''|\partial_r\epsilon|^2 + \frac{1}{2}\int \mathbf{1}_{r>A}\frac{\phi_A'}{r}\Big(\frac{m+A_\theta[Q]}{r}\Big)^2|\epsilon|^2 \\ &- \frac{1}{8}\int(\Delta^2\phi_A)|\epsilon|^2 + (\mathrm{bdry}), \end{aligned}$$

where (bdry) is a boundary term evaluated at $r = A$. The first and second lines are apparently nonnegative. For the third line, a crude estimate

$$\Big|\frac{1}{8}\int(\Delta^2\phi_A)|\epsilon|^2\Big| \lesssim \|r^{-1}\epsilon\|_{L^2}^2$$

using $|\Delta^2\phi_A| \lesssim \mathbf{1}_{r\geq A}\frac{A}{r^3}$ is on the borderline of acceptable error size. So is (bdry). More precisely, if ϵ is localized on the region $\{r \sim A\}$, we would not have an improvement. One of the reasons causing this difficulty is that the coercivity of $\lambda^2\mathcal{I}_A$ *does not* control L^2 norm of ϵ (uniformly in η). This is a difference from (NLS). To resolve this, we average $\lambda^2\mathcal{I}_A$ as

$$\lambda^2\mathcal{I} = \frac{2}{\log A}\int_{A^{1/2}}^{A}\lambda^2\mathcal{I}_{A'}\frac{dA'}{A'}.$$

Then, we can prove that $\lambda^2\partial_s\mathcal{I}$ is almost nonnegative. With this $\mathcal{I}$, we can close our bootstrap procedure. See Figure 1.

7. *Conditional uniqueness.* Let u_1 and u_2 be as in Theorem 1.2. We decompose each u_j with its modulation parameters $(b_j, \lambda_j, \gamma_j)$ and ϵ_j, and redo the modulation analysis on $\epsilon := \epsilon_1 - \epsilon_2$. A similar argument is used in [**39**]. We conclude $\epsilon = 0$ and

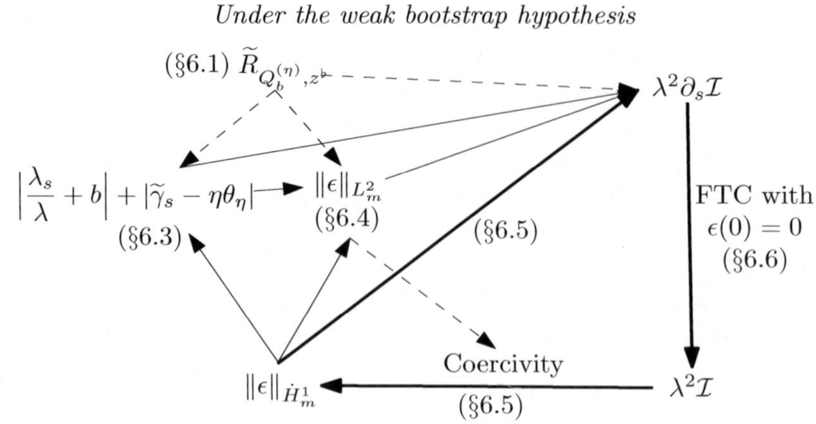

FIGURE 1. Scheme of the bootstrap

$(b_1,\lambda_1,\gamma_1)=(b_2,\lambda_2,\gamma_2)$ again by Lyapunov/virial method. Here the crucial idea is to write for $j\in\{1,2\}$

$$u_j(t,x)=\frac{e^{i\gamma_j}}{\lambda_j}Q_{b_j}\Big(\frac{1}{\lambda_j}\Big)+z(t,x)+\frac{e^{i\gamma_1}}{\lambda_1}\epsilon_j(t,\frac{x}{\lambda_1}).$$

We take the $\flat$-operation with the common parameter set (λ_1,γ_1). Here, the point is that u_1 and u_2 should be rescaled with the common parameters since the hypothesis is $\|u_1(t)-u_2(t)\|_{H^1}\leq c|t|$. From this, the parameter difference only appears in the difference of $Q_{b_1,\lambda_1,\gamma_2}$ and $Q_{b_2,\lambda_2,\gamma_2}$. This leads that estimating ϵ boils down to estimating the parameter differences. Furthermore, we observe that (1.39) with $\eta=0$ is explicitly solvable:

$$\frac{b_1}{(\lambda_1)^2}=|t|^{-1}=\frac{b_2}{(\lambda_2)^2}.$$

This observation improves the estimate of $Q_{b_1,\lambda_1,\gamma_2}$ and $Q_{b_2,\lambda_2,\gamma_2}$ by a logarithmic factor, which is crucial to conclude the proof.

Organization of the paper

Throughout the paper, we present analysis with the η-parameter, $u^{(\eta)}$. The reader who is interested only in the construction (or, anyone at first reading) may read all assuming $\eta=0$ with a minor correction. In fact, the authors conducted this work with $\eta=0$ first.

In Chapter 2, we collect notations, some properties of equivariant Sobolev spaces, and preliminary multilinear estimates. We also introduce time maximal functions on $[t_0^*,0)$ and present equivariant Cauchy theory on L^2_m and H^s_m. In Chapter 3, we discuss the linearization of (CSS). In Chapter 4, we discuss an instability mechanism and construct the profile $Q^{(\eta)}$. In Chapter 5, we set up the modulation analysis, fix modulation parameters, and reduce Theorems 1.1 and 1.3 to the main bootstrap lemma. In Chapter 6, we close our bootstrap argument by employing the Lyapunov/virial functional method. In Chapter 7, we prove Theorem 1.2.

There are two appendices. In Appendix A, we discuss equivariant functions and Sobolev spaces for self-containedness. In Appendix B, we sketch the proof of equivariant H^s_m-Cauchy theory (Proposition 2.10).

Acknowledgments

We appreciate A. Lawrie, S.-J. Oh, and S. Shahshahani for sharing their manuscript [**33**] on the linearization of (CSS). We are also grateful to S.-J. Oh for helpful discussions and encouragement to this work.

CHAPTER 2

Notations and preliminaries

2.1. Basic notations

We mainly work with equivariant (including radial) functions on $\mathbb{R}^2$, say $\phi : \mathbb{R}^2 \to \mathbb{C}$. We also work with their radial part $u : (0,\infty) \to \mathbb{C}$. Thus we use the integral symbol $\int$ to mean

$$\int = \int_{\mathbb{R}^2} dx = 2\pi \int_0^\infty r dr. \tag{2.1}$$

We also need the generator Λ for the L^2-scaling, which is defined by

$$\Lambda f := \frac{d}{d\lambda}\Big|_{\lambda=1} \lambda f(\lambda \cdot) = [1 + r\partial_r] f. \tag{2.2}$$

The linearized operator at the static solution is only $\mathbb{R}$-linear. We thus work with $L^2(\mathbb{R}^2;\mathbb{C})$ viewed as a *real* Hilbert space with the inner product

$$(f,g)_r := \int \mathrm{Re}(f\overline{g}). \tag{2.3}$$

All the functional derivatives will be computed with respect to this real inner product.

For $\eta \geq 0$, we will use the *η-dependent Japanese bracket*:

$$\langle t \rangle := (t^2 + \eta^2)^{\frac{1}{2}}. \tag{2.4}$$

In addition, we will use notations such as L^2_m, $\dot{H}^1_m$, $f^\sharp$, $g^\flat$, f_b, and so on. These will be defined on the way.

We also use the conventional notations. We say $A \lesssim B$ when there is some implicit constant C that does not depend on A and B satisfying $A \leq CB$. For some parameter r, we write $A \lesssim_r B$ if the implicit constant C depends on r. In this paper, dependence on the equivariance index m is suppressed; we simply write $A \lesssim B$ for $A \lesssim_m B$.

We use (mixed) Lebesgue norms. For $1 \leq p \leq \infty$, we define ℓ^p and L^p norms by

$$\|a\|_{\ell^p} := \Big(\sum_{k\geq 0} |a_k|^p \Big)^{\frac{1}{p}},$$

$$\|f\|_{L^p} := \Big(\int_{\mathbb{R}^d} |f(x)|^p dx \Big)^{\frac{1}{p}},$$

for any sequence $a = (a_0, a_1, \dots)$ and function $f : \mathbb{R}^d \to \mathbb{C}$. For $1 \leq q \leq \infty$, an interval $I \subseteq \mathbb{R}$, a Banach space X, and $f : I \to X$, we denote

$$\|f\|_{L^q_I X} := \big\| \|f(t)\|_X \big\|_{L^q_t(I)}.$$

In particular, we simply write $L_I^q := L_t^q(I)$. In case of $I = \mathbb{R}$, we write $L_t^q := L_t^q(\mathbb{R})$. If $t \in I \mapsto G(t) \in X$ is continuous and bounded, we write $G \in C_I X$. We also use $f \in L_{I,\mathrm{loc}}^q X$ or $f \in C_{I,\mathrm{loc}} X$ if the restriction of f on any subintervals J of I belong to $L_J^q X$ or $C_J X$, respectively. Finally, the mixed Lebesgue norm $L_I^q L_x^q$ is simply written as $L_{I,x}^q$ in view of Fubini's theorem.

We also need the following Sobolev norms

$$\|f\|_{\dot{H}^1} := \|\nabla f\|_{L^2} \quad \text{and} \quad \|f\|_{H^1} := (\|f\|_{L^2}^2 + \|f\|_{\dot{H}^1}^2)^{\frac{1}{2}}.$$

In some places (but not often), we need fractional Sobolev norms ($s \in \mathbb{R}$)

$$\|f\|_{H^s} := \|\langle D\rangle^s f\|_{L^2},$$

where $\langle D\rangle^s$ is the Fourier multiplier operator with the symbol $\langle \xi\rangle^s$.

2.2. Equivariant Sobolev spaces

Let $m \in \mathbb{Z} \setminus \{0\}$. In this section, we record some properties of m-equivariant functions and associated Sobolev spaces. We confine to $m \neq 0$ for simplicity of the exposition. We also assume $m > 0$ since m-equivariant functions with $m < 0$ are merely conjugates of $(-m)$-equivariant function. An elementary detailed exposition is given in Appendix A.

A function $f : \mathbb{R}^2 \to \mathbb{C}$ is said to be *m-equivariant* if

$$f(x) = g(r)e^{im\theta}, \qquad \forall x \in \mathbb{R}^2 \setminus \{0\}$$

for some $g : (0,\infty) \to \mathbb{C}$, where we expressed $x_1 + ix_2 = re^{i\theta}$ with $x = (x_1, x_2)$.

Here is a brief discussion of smooth m-equivariant functions f for $m \neq 0$. If we require f to be continuous at the origin, we should have $f(0) = 0$ because the factor $e^{im\theta}$ is not continuous at the origin. Moreover, regularity of f forces f to be degenerate at the origin; in fact, Lemma A.3 says that smooth m-equivariant f should have the expression[1]

$$f(x) = h(x) \cdot (x_1 + ix_2)^m$$

for some *smooth* radial function $h(x) = h(|x|)$. This explicitly shows the degeneracy of f at the origin from its regularity and m.

We turn our attention to equivariant Sobolev spaces. For $s \geq 0$, we define H_m^s by the set of m-equivariant functions lying in the usual Sobolev space $H^s(\mathbb{R}^2)$ with denoting $L_m^2 := H_m^0$. We equip H_m^s with the usual $H^s(\mathbb{R}^2)$-norm. We denote by $C_{c,m}^\infty(\mathbb{R}^2 \setminus \{0\})$ the set of smooth m-equivariant functions having compact support in $\mathbb{R}^2 \setminus \{0\}$. Here is a useful density theorem for equivariant functions (Lemma A.5): the space $C_{c,m}^\infty(\mathbb{R}^2 \setminus \{0\})$ is dense in H_m^s if and only if $s \leq m + 1$.

For functions in $C_{c,m}^\infty(\mathbb{R}^2 \setminus \{0\})$, we have

$$\begin{aligned} \Delta f &= \partial_{rr} f + \frac{1}{r}\partial_r f - \frac{m^2}{r^2} f, \\ \|f\|_{\dot{H}^1}^2 &= \|\nabla f\|_{L^2}^2 = \|\partial_r f\|_{L^2}^2 + m^2\|r^{-1} f\|_{L^2}^2, \\ \|f\|_{H^1}^2 &= \|f\|_{L^2}^2 + \|f\|_{\dot{H}^1}^2. \end{aligned}$$

In particular, we have the Hardy-Sobolev inequality (Lemma A.6): for $m \neq 0$,

$$\|r^{-1} f\|_{L^2} + \|f\|_{L^\infty} \lesssim \|f\|_{\dot{H}^1}, \qquad \forall f \in C_{c,m}^\infty(\mathbb{R}^2 \setminus \{0\}). \tag{2.5}$$

[1] If $m < 0$, then f has expression $f(x) = h(x) \cdot (x_1 - ix_2)^{|m|}$.

There is also a generalization of Hardy's inequality (Lemma A.7): if $0 \le k \le m$ is an integer, then we have
(2.6)

$$\|r^{-k} f\|_{L^2} + \|r^{-(k-1)}\partial_r f\|_{L^2} + \cdots + \|\partial_r^k f\|_{L^2} \sim \|f\|_{\dot{H}^k}, \qquad \forall f \in C^\infty_{c,m}(\mathbb{R}^2 \setminus \{0\}).$$

Moreover, one can also have L^∞-estimate for derivatives of f; for example, we have

$$\|\partial_r f\|_{L^\infty} \lesssim \|f\|_{\dot{H}^2}, \qquad \forall m \ge 2,\ f \in C^\infty_{c,m}(\mathbb{R}^2 \setminus \{0\}) \tag{2.7}$$

by mimicking the proof of (2.5).

For an integer $0 \le k \le m$, we define the space $\dot{H}^k_m$ by the completion of $C^\infty_{c,m}(\mathbb{R}^2 \setminus \{0\})$ in $L^2_{\text{loc}}(\mathbb{R}^2)$ with respect to $\dot{H}^k_m$ norm. We have the following embeddings of spaces:

$$C^\infty_{c,m}(\mathbb{R}^2 \setminus \{0\}) \hookrightarrow H^k_m \hookrightarrow \dot{H}^k_m \hookrightarrow L^2_{\text{loc}}(\mathbb{R}^2).$$

By density, it suffices to check estimates regarding H^k_m or $\dot{H}^k_m$ norms for $C^\infty_{c,m}(\mathbb{R}^2 \setminus \{0\})$ functions. In particular, (2.5), (2.6), and (2.7) hold for all functions in $\dot{H}^1_m$, $\dot{H}^k_m$, and $\dot{H}^2_m$, respectively. In this paper, the space $\dot{H}^1_m$ is of particular importance.

2.3. Time maximal functions

In this section, we introduce time maximal functions with the time variable t. In the proof of Theorems 1.1-1.3, we will use these for functions defined on $[t_0^*, 0)$. Several estimates (say at time $t < 0$) will be obtained by a time integral on $[t, 0)$. For example, $\|\epsilon(t)\|$ will be estimated by a time integral of $\epsilon(t')$ on $[t, 0)$ with $\epsilon(0) = 0$. Thus it is convenient to use a maximal function of time instead of fixed-time quantities.

To motivate this, consider the following situation. If $f : [t_0^*, 0) \to \mathbb{R}$ is given by $f(t) \sim |t|^p$, $p \ge 0$, then we have $\int_t^0 |f(t')|dt' \sim \frac{|t|^{p+1}}{p+1} \sim_p |t||f(t)|$. However, if we only know $f(t) \lesssim |t|^p$, then we cannot guarantee that $\int_t^0 |f(t')|dt' \lesssim_p |t||f(t)|$. Instead, we want to introduce some maximal function $\mathcal{T}f$ such that $\int_t^0 |f(t')|dt' \lesssim_p |t|[\mathcal{T}f](t)$.

Fix $\eta \ge 0$. We consider a function $f : [t_0^*, 0) \to X$, where X is a Banach space and $t_0^* \in (-\infty, 0)$. For $t \in [t_0^*, 0)$, let us introduce

$$n(t) := \begin{cases} \lceil \max\{0, \log_2 \frac{|t|}{\eta}\} \rceil, & \text{if } \eta > 0, \\ \infty & \text{if } \eta = 0, \end{cases}$$

where $\lceil \alpha \rceil$ is the smallest integer greater than or equal to α. For each nonnegative integer j, we define

$$a[f; t, j] := \begin{cases} \sup_{t' \in [2^{-j}t, 2^{-(j+1)}t)} \|f(t')\|_X & \text{if } 0 \le j < n(t), \\ \sup_{t' \in [2^{-j}t, 0)} \|f(t')\|_X & \text{if } j \ge n(t). \end{cases}$$

For each $s \in \mathbb{R}$, we define the *time maximal function* $\mathcal{T}^{(\eta,s)}_X[f] : [t_0^*, 0) \to [0, +\infty]$ by

$$\mathcal{T}^{(\eta,s)}_X[f](t) := \sum_{j=0}^{n(t)} 2^{js}(j^2+1)^{-1} a[f; t, j]. \tag{2.8}$$

When $\eta = 0$, we denote $\mathcal{T}_X^{(0,s)}[f] = \mathcal{T}_X^{(s)}[f]$. When f is scalar-valued, we simply write $\mathcal{T}_X^{(\eta,s)}[f] = \mathcal{T}^{(\eta,s)}[f]$. The decay rate $(j^2+1)^{-1}$ in (2.8) is a technical factor that is required in the proof of Lemma 2.1 (see (2.15) and (2.17)).

It is clear that $\mathcal{T}_X^{(\eta,s)}$ is sub-additive. Indeed, for any scalars c_n and functions f_n, we have

$$\mathcal{T}_X^{(\eta,s)}\Big[\sum_n c_n f_n\Big] \leq \sum_n |c_n| \mathcal{T}_X^{(\eta,s)}[f_n].$$

$\mathcal{T}_X^{(\eta,s)}$ preserves power-type bounds; if $\|f(t)\|_X \sim \langle t\rangle^q$ and $s \leq q$ (recall that we defined $\langle t\rangle := (|t|^2+\eta^2)^{\frac{1}{2}}$), then we have

$$\mathcal{T}_X^{(\eta,s)}[f](t) \sim_q \langle t\rangle^q.$$

$\mathcal{T}_X^{(\eta,s)}$ is idempotent in the sense of (2.10). Moreover, $\mathcal{T}_X^{(\eta,s)}$ is designed to behave nicely with integration in t; see (2.12) below.

One can easily verify qualitative properties of $\mathcal{T}_X^{(\eta,s)}$. If $\eta > 0$ or $f(t) \lesssim |t|^s$ for t near zero, then $\mathcal{T}_X^{(\eta,s)}[f]$ has finite values. If $\mathcal{T}_X^{(\eta,s)}[f](t_0) < \infty$ for some $t_0 \in [t_0^*, 0)$, then $\mathcal{T}_X^{(\eta,s)}[f](t) < \infty$ for all $t \in [t_0, 0)$. Finally, if f is continuous, then so is $\mathcal{T}_X^{(\eta,s)}[f]$.

We conclude this section with quantitative properties.

LEMMA 2.1 (Quantitative properties of $\mathcal{T}_X^{(\eta,s)}$). *Fix $\eta \geq 0$. Let $q \in \mathbb{R}$, $s \geq 0$, and $1 \leq p \leq \infty$. Then,*

$$\|f(t)\|_X \leq \mathcal{T}_X^{(\eta,s)}[f](t), \tag{2.9}$$

$$\mathcal{T}^{(\eta,s)}[\mathcal{T}_X^{(\eta,s)}[f]](t) \sim_s \mathcal{T}_X^{(\eta,s)}[f](t), \tag{2.10}$$

$$\mathcal{T}_X^{(\eta,s)}[\langle\cdot\rangle^q f](t) \sim_q \langle t\rangle^q \mathcal{T}_X^{(\eta,s-q)}[f](t). \tag{2.11}$$

If $\frac{1}{p}+q+s > 0$, we have

$$\|\langle\cdot\rangle^q \mathcal{T}_X^{(\eta,s)}[f]\|_{L^p_{[t,0)}} \lesssim_{p,q,s} \langle t\rangle^{q+\frac{1}{p}} \mathcal{T}_X^{(\eta,s)}[f](t). \tag{2.12}$$

We remark that all of the above estimates are uniform in $\eta \geq 0$.

PROOF. (2.9) is immediate from the definition (2.8).

We turn to (2.10). We have the ($\gtrsim$) direction by (2.9). To show the ($\lesssim$) direction, we use the following properties of $a[f;t,j]$:

$$a[a[f;\cdot,k];t,j] \leq a[f;t,j+k] + a[f;t,j+k+1] \text{ for all } j,k \geq 0, \tag{2.13}$$

$$j+k \leq n(t) \text{ if } 0 \leq j \leq n(t) \text{ and } 0 \leq k \leq n(2^{-j}t). \tag{2.14}$$

By (2.13) and (2.14), we have

$$\begin{aligned}
&\mathcal{T}^{(\eta,s)}[\mathcal{T}_X^{(\eta,s)}[f]](t) \\
&\leq \sum_{j=0}^{n(t)} \sum_{k=0}^{n(2^{-j}t)} 2^{(j+k)s}(j^2+1)^{-1}(k^2+1)^{-1} a[a[f;\cdot,k];t,j] \\
&\leq \sum_{\ell=0}^{n(t)} \Big(\sum_{j=0}^{\ell} \frac{\ell^2+1}{(j^2+1)((\ell-j)^2+1)} \Big) 2^{\ell s}(\ell^2+1)^{-1}\Big(a[f;t,\ell] + a[f;t,\ell+1]\Big).
\end{aligned}$$

Using

$$\sum_{j=0}^{\ell} \frac{\ell^2+1}{(j^2+1)((\ell-j)^2+1)} \lesssim \sum_{j=0}^{\infty} \frac{1}{j^2+1} \lesssim 1, \tag{2.15}$$

the estimate (2.10) follows.

The estimate (2.11) easily follows from the observation

$$\langle t'\rangle \sim 2^{-j}\langle t\rangle \quad \text{whenever} \begin{cases} t' \in [2^{-j}t, 2^{-(j+1)}t) & \text{if } j < n(t), \text{ or} \\ t' \in [2^{-j}t, 0) & \text{if } j = n(t). \end{cases} \tag{2.16}$$

Finally, we assume $\frac{1}{p}+q+s>0$ and show (2.12). We only provide the proof for $1 \le p < \infty$ as the case $p=\infty$ can be proved with an obvious modification of the argument. We estimate the LHS of (2.12) as

$$\begin{aligned} &\|\langle \cdot\rangle^q \mathcal{T}_X^{(\eta,s)}[f]\|_{L^p_{[t,0)}} \\ &\lesssim \||2^{-j}t|^{\frac{1}{p}} \langle 2^{-j}t\rangle^q \mathcal{T}_X^{(\eta,s)}[f](2^{-j}t)\|_{\ell^p_{0\le j\le n(t)}} \\ &\lesssim_q |t|^{\frac{1}{p}}\langle t\rangle^q \big\|2^{-j(\frac{1}{p}+q)} \cdot 2^{ks}(k^2+1)^{-1}a[f;2^{-j}t,k]\big\|_{\ell^p_{0\le j\le n(t)}\ell^1_{0\le k\le n(2^{-j}t)}}, \end{aligned}$$

where we used (2.16) in the last inequality. We then estimate the above $\ell^p_j\ell^1_k$ term using Hölder's inequality with

$$\begin{aligned} \big\|2^{-j(\frac{1}{p}+q+s)}((j+k)^2+1)(k^2+1)^{-2}\big\|_{\ell^p_{j\ge0}\ell^\infty_{k\ge0}} &\lesssim_{p,q,s} 1, \\ \big\|2^{(j+k)s}((j+k)^2+1)^{-1}a[f;t,j+k]\big\|_{\ell^\infty_{0\le j\le n(t)}\ell^1_{0\le k\le n(2^{-j}t)}} &\le \mathcal{T}_X^{(\eta,s)}[f](t), \end{aligned} \tag{2.17}$$

where we used $a[f;2^{-j}t,k] = a[f;t,j+k]$ and (2.14). This completes the proof of (2.12). □

2.4. Decomposition of nonlinearity and duality estimates

Decomposition of nonlinearity. Denote by $\mathcal{N}(u)$ the nonlinearity of (CSS):

$$\mathcal{N}(u) := -|u|^2u + \frac{2m}{r^2}A_\theta[u]u + \frac{A_\theta^2[u]}{r^2}u + A_0[u]u. \tag{2.18}$$

We introduce trilinear terms $\mathcal{N}_{3,0}$, $\mathcal{N}_{3,1}$, $\mathcal{N}_{3,2}$ and quintilinear terms $\mathcal{N}_{5,1}$, $\mathcal{N}_{5,2}$. For $\psi_j : (0,\infty) \to \mathbb{C}$, with abbreviations

$$\begin{aligned} \mathcal{N}_{3,k} &= \mathcal{N}_{3,k}(\psi_1,\psi_2,\psi_3), \quad \forall k \in \{0,1,2\} \\ \mathcal{N}_{5,k} &= \mathcal{N}_{5,k}(\psi_1,\psi_2,\psi_3,\psi_4,\psi_5), \quad \forall k \in \{1,2\}, \end{aligned}$$

we define

$$
\begin{cases}
\mathcal{N}_{3,0} := -\psi_1\overline{\psi_2}\psi_3, \\
\mathcal{N}_{3,1} := -\frac{m}{r^2}\Big(\int_0^r \mathrm{Re}(\psi_1\overline{\psi_2})r'dr'\Big)\psi_3, \\
\mathcal{N}_{3,2} := -\Big(\int_r^\infty \frac{m}{r'}\mathrm{Re}(\psi_1\overline{\psi_2})dr'\Big)\psi_3, \\
\mathcal{N}_{5,1} := \frac{1}{4r^2}\Big(\int_0^r \mathrm{Re}(\psi_1\overline{\psi_2})r'dr'\Big)\Big(\int_0^r \mathrm{Re}(\psi_3\overline{\psi_4})r'dr'\Big)\psi_5, \\
\mathcal{N}_{5,2} := \frac{1}{2}\Big(\int_r^\infty \Big(\int_0^{r'} \mathrm{Re}(\psi_1\overline{\psi_2})r''dr''\Big)\mathrm{Re}(\psi_3\overline{\psi_4})\frac{dr'}{r'}\Big)\psi_5.
\end{cases} \tag{2.19}
$$

In case of $\psi_j = u$ for all j, we write

$$\mathcal{N}_*(u) := \mathcal{N}_*(u, \dots, u)$$

for any possible choice of $*$. Then we can write the nonlinearity $\mathcal{N}$ as

$$\mathcal{N}(u) = [\mathcal{N}_{3,0} + \mathcal{N}_{3,1} + \mathcal{N}_{3,2} + \mathcal{N}_{5,1} + \mathcal{N}_{5,2}](u).$$

Introduction of $\mathcal{M}_{4,0}^{(A)}$, $\mathcal{M}_{4,1}^{(A)}$, and $\mathcal{M}_6^{(A)}$. We introduce a smooth radial weight ϕ such that $\partial_r\phi$ is increasing and

$$\partial_r\phi(r) = \begin{cases} r & \text{if } r \le 1, \\ 3 - e^{-r} & \text{if } r \ge 2. \end{cases}$$

We write for $A \in [1, +\infty]$:

$$\phi_A(r) := \begin{cases} A^2\phi(\frac{r}{A}), & \text{if } 1 \le A < \infty, \\ \frac{1}{2}r^2 & \text{if } A = \infty. \end{cases} \tag{2.20}$$

We introduce quartic forms $\mathcal{M}_{4,0}^{(A)}$, $\mathcal{M}_{4,1}^{(A)}$, and sextic form $\mathcal{M}_6^{(A)}$. For $\psi_j : (0,\infty) \to \mathbb{C}$, with abbreviations

$$
\begin{aligned}
\mathcal{M}_{4,k}^{(A)} &= \mathcal{M}_{4,k}^{(A)}(\psi_1, \psi_2, \psi_3, \psi_4), \quad \forall k \in \{0, 1\} \\
\mathcal{M}_6^{(A)} &= \mathcal{M}_6^{(A)}(\psi_1, \psi_2, \psi_3, \psi_4, \psi_5, \psi_6),
\end{aligned}
$$

we define

$$
\begin{cases}
\mathcal{M}_{4,0}^{(A)} := \int \frac{\Delta\phi_A}{2} \cdot \psi_1\overline{\psi_2}\psi_3\overline{\psi_4}, \\
\mathcal{M}_{4,1}^{(A)} := \int \frac{\phi_A'}{r} \cdot \frac{1}{r^2}\Big(\int_0^r \mathrm{Re}(\psi_1\overline{\psi_2})r'dr'\Big)\mathrm{Re}(\psi_3\overline{\psi_4}), \\
\mathcal{M}_6^{(A)} := \int \frac{\phi_A'}{r} \cdot \frac{1}{r^2}\Big(\int_0^r \mathrm{Re}(\psi_1\overline{\psi_2})r'dr'\Big)\Big(\int_0^r \mathrm{Re}(\psi_3\overline{\psi_4})r'dr'\Big)\mathrm{Re}(\psi_5\overline{\psi_6}).
\end{cases} \tag{2.21}
$$

In case of $\psi_k = u$ for all k, we simply write as

$$\mathcal{M}_*^{(A)}(u) := \mathcal{M}_*^{(A)}(u, \dots, u)$$

for any choice of $*$. In case of $A = \infty$, we write as

$$\mathcal{M}_* := \mathcal{M}_*^{(\infty)}.$$

We will mostly use the case $A = \infty$; the case $1 \le A < \infty$ will only appear when we do analysis with the Lyapunov/virial functional in Section 6.5.

The above forms $\mathcal{M}_*$ arise naturally in the energy functional. In particular, we can rewrite the energy functional $E[u]$ as

$$E[u] = \frac{1}{2}\int \Big(|\partial_r u|^2 + \frac{m^2}{r^2}|u|^2\Big) - \frac{1}{4}\mathcal{M}_{4,0}(u) - \frac{m}{2}\mathcal{M}_{4,1}(u) + \frac{1}{8}\mathcal{M}_6(u). \tag{2.22}$$

If we recall that (CSS) has a Hamiltonian formulation with the Hamiltonian E, the nonlinearity $\mathcal{N}$ arises in the functional derivative of the energy. We in fact have

$$\begin{cases} (\mathcal{N}_{3,0}(\psi_1,\psi_2,\psi_3),\psi_4)_r = \mathcal{M}_{4,0}(\psi_1,\psi_2,\psi_3,\psi_4), \\ (\mathcal{N}_{3,1}(\psi_1,\psi_2,\psi_3),\psi_4)_r = -m\mathcal{M}_{4,1}(\psi_1,\psi_2,\psi_3,\psi_4), \\ (\mathcal{N}_{3,2}(\psi_1,\psi_2,\psi_3),\psi_4)_r = -m\mathcal{M}_{4,1}(\psi_3,\psi_4,\psi_1,\psi_2), \\ (\mathcal{N}_{5,1}(\psi_1,\psi_2,\psi_3,\psi_4,\psi_5),\psi_6)_r = \frac{1}{4}\mathcal{M}_6(\psi_1,\psi_2,\psi_3,\psi_4,\psi_5,\psi_6), \\ (\mathcal{N}_{5,2}(\psi_1,\psi_2,\psi_3,\psi_4,\psi_5),\psi_6)_r = \frac{1}{2}\mathcal{M}_6(\psi_1,\psi_2,\psi_5,\psi_6,\psi_3,\psi_4). \end{cases} \tag{2.23}$$

In view of duality, this says that estimates of multilinear forms $\mathcal{M}_*$ transfer to estimates of $\mathcal{N}_*$. This formulation simplifies multilinear estimates of $\mathcal{N}_*$.

We conclude this section by recording some duality estimates. The first one is well-adapted to the fixed time energy estimate.

LEMMA 2.2 (Duality estimates, I). *Let $A \in [1,\infty]$. The following estimates hold uniformly in A.*

(1) *For any distinct indices $i, i' \in \{1,2,3,4\}$, we have*

$$|\mathcal{M}_{4,0}^{(A)}| \lesssim \|\psi_i\|_{L^\infty}\|\psi_{i'}\|_{L^\infty}\prod_{j\neq i,i'}\|\psi_j\|_{L^2}. \tag{2.24}$$

(2) *For any distinct indices $i, i' \in \{1,2,3,4\}$, we have*

$$|\mathcal{M}_{4,1}^{(A)}| \lesssim \Big\|\frac{\psi_i}{r}\Big\|_{L^2}\Big\|\frac{\psi_{i'}}{r}\Big\|_{L^2}\prod_{j\neq i,i'}\|\psi_j\|_{L^2}. \tag{2.25}$$

(3) *For any distinct indices $i, i' \in \{1,\dots,6\}$, we have*

$$|\mathcal{M}_{6}^{(A)}| \lesssim \Big\|\frac{\psi_i}{r}\Big\|_{L^2}\Big\|\frac{\psi_{i'}}{r}\Big\|_{L^2}\prod_{j\neq i,i'}\|\psi_j\|_{L^2}. \tag{2.26}$$

PROOF. In each case, it suffices to consider the case with $A = \infty$ since we have $|\Delta\phi_A| + |\frac{\phi_A'}{r}| \lesssim 1$ uniformly in A. The estimate (2.24) easily follows by Hölder's inequality. To show (2.25) and (2.26), decompose $\frac{1}{r^2} = \frac{1}{r}\cdot\frac{1}{r}$ and move each $\frac{1}{r}$ to ψ_i and $\psi_{i'}$. We note that one can put $\frac{1}{r}$ inside the integral as

$$\Big|\frac{1}{r}\int_0^r \mathrm{Re}(\overline{\psi}\varphi)r'dr'\Big| \lesssim \int_0^r \Big|\frac{\psi}{r'}\cdot\varphi\Big|r'dr'.$$

Now the estimates (2.25) and (2.26) follow by Hölder's inequality with

$$\Big\|\int_0^r \mathrm{Re}(\overline{\psi}\varphi)r'dr'\Big\|_{L^\infty} \lesssim \|\psi\|_{L^2}\|\varphi\|_{L^2}.$$

This completes the proof. □

LEMMA 2.3 (H^1_m-estimate of $\mathcal{N}$). *Let $m \geq 1$. For any possible choice of $*$ in (2.19) and any two distinct indices i, i', we have*

$$\|\mathcal{N}_*\|_{L^2_m} \lesssim \|\psi_i\|_{\dot{H}^1_m} \|\psi_{i'}\|_{\dot{H}^1_m} \prod_{j \neq i,i'} \|\psi_j\|_{L^2_m}, \tag{2.27}$$

$$\|\mathcal{N}_*\|_{H^1_m} \lesssim \|\psi_i\|_{\dot{H}^1_m} \|\psi_{i'}\|_{\dot{H}^1_m} \prod_{j \neq i,i'} \|\psi_j\|_{H^1_m}. \tag{2.28}$$

PROOF. The estimate (2.27) easily follows from the duality relations (2.23), duality estimates (Lemma 2.2), and embeddings (2.5).

To estimate (2.28), we act ∂_r and $\frac{1}{r}$ on $\mathcal{N}_{3,0}, \mathcal{N}_{3,1}, \mathcal{N}_{3,2}$. These consist of linear combinations of the following expressions:

$$(\partial_r\psi_1)\overline{\psi_2}\psi_3; \quad \psi_1(\partial_r\overline{\psi_2})\psi_3; \quad \psi_1\overline{\psi_2}(\partial_r\psi_3); \quad \psi_1\overline{\psi_2}\tfrac{\psi_3}{r};$$
$$(\int_r^\infty \psi_1\overline{\psi_2}\frac{dr'}{r'})\frac{\psi_3}{r}; \quad \frac{1}{r^2}(\int_0^r \psi_1\overline{\psi_2}r'dr')\frac{\psi_3}{r}.$$

These expressions are of types $\mathcal{N}_{3,0}$, $\mathcal{N}_{3,1}$, and $\mathcal{N}_{3,2}$, where we replace exactly one ψ_j by $\partial_r\psi_j$ or $r^{-1}\psi_j$. Thus we can apply (2.27) and (2.5) to estimate all the above terms by

$$\|\psi_1\|_{\dot{H}^1_m} \|\psi_2\|_{\dot{H}^1_m} \|\psi_3\|_{\dot{H}^1_m}.$$

We turn to $\mathcal{N}_{5,1}$ and $\mathcal{N}_{5,2}$. As before, we will investigate what expressions arise when ∂_r and $\frac{1}{r}$ act on $\mathcal{N}_{5,1}$ and $\mathcal{N}_{5,2}$. We first note the following useful tricks:

$$\partial_r\Big(\int_0^r \psi_1\overline{\psi_2}r'dr'\Big) = \int_0^r \Big((\partial_r\psi_1)\overline{\psi_2} + \psi_1(\partial_r\overline{\psi_2}) + \frac{1}{r'}\psi_1\overline{\psi_2}\Big)r'dr',$$
$$\partial_r\Big(\int_r^\infty \psi_1\overline{\psi_2}\frac{dr'}{r'}\Big) = \int_r^\infty \Big((\partial_r\psi_1)\overline{\psi_2} + \psi_1(\partial_r\overline{\psi_2}) - \frac{1}{r'}\psi_1\overline{\psi_2}\Big)\frac{dr'}{r'}.$$

Therefore, ∂_r and $\frac{1}{r}$ acted on $\mathcal{N}_{5,1}$ and $\mathcal{N}_{5,2}$ consist of expressions of types $\mathcal{N}_{5,1}$ and $\mathcal{N}_{5,2}$, where we replace exactly one ψ_j by $\partial_r\psi_j$ or $r^{-1}\psi_j$. Applying (2.27) completes the proof. □

We also need the following duality estimate adapted to the Strichartz estimates. It will be used mostly in Section 6.4 in a crucial way.

LEMMA 2.4 (Duality estimates, II). *Let $A \in [1, \infty]$. The following estimates hold uniformly in A.*

(1) *For any distinct indices $i, i' \in \{1, \dots, 4\}$, we have*

$$|\mathcal{M}^{(A)}_{4,0}| \lesssim \|\psi_i\|_{L^\infty} \|\psi_{i'}\|_{L^4} \prod_{j \neq i,i'} \|\psi_j\|_{L^{p_j}} \tag{2.29}$$

for any $p_j \in \{2, 4\}$ $(j \neq i, i')$ satisfying (2.32).

(2) *For any distinct indices $i, i' \in \{1, \dots, 4\}$, there exist $p_j \in \{2, 4\}$ $(j \neq i, i')$ such that*

$$|\mathcal{M}^{(A)}_{4,1}| \lesssim \Big\|\frac{\psi_i}{r}\Big\|_{L^2} \|\psi_{i'}\|_{L^4} \prod_{j \neq i,i'} \|\psi_j\|_{L^{p_j}}. \tag{2.30}$$

(3) *For any distinct indices $i, i' \in \{1, \dots, 6\}$, there exist $p_j \in \{2, 4\}$ $(j \neq i, i')$ such that*

$$|\mathcal{M}^{(A)}_{6}| \lesssim \Big\|\frac{\psi_i}{r}\Big\|_{L^2} \|\psi_{i'}\|_{L^4} \prod_{j \neq i,i'} \|\psi_j\|_{L^{p_j}}. \tag{2.31}$$

In the above estimates, p_j's satisfy the scaling conditions

$$\sum_{j\neq i,i'} \frac{1}{p_j} = \begin{cases} \frac{3}{4} & \textit{for } (2.29) \textit{ and } (2.30), \\ \frac{7}{4} & \textit{for } (2.31). \end{cases} \tag{2.32}$$

REMARK 2.5. Due to expressions of $\mathcal{M}_{4,1}$ and $\mathcal{M}_6$, we cannot choose arbitrary p_j's satisfying the scaling condition (2.32). The choice of p_j's *depends* on the pair (i,i').

PROOF OF LEMMA 2.4. In each case, it suffices to consider the case with $A=\infty$ since we have $|\Delta\phi_A|+|\frac{\phi_A'}{r}|\lesssim 1$ uniformly in A. The estimate (2.29) clearly follows by Hölder's inequality. We omit the proof of (2.30) and only show (2.31), as the former is easier than the latter. Henceforth, we focus on (2.31) with $A=\infty$. Due to symmetry, it suffices to consider five cases: $(i,i')\in\{(1,2),(1,3),(1,5),(5,1),(5,6)\}$. We will rely on

$$\Big\|\frac{1}{r}\int_0^r |f|r'dr'\Big\|_{L^4}\lesssim \|f\|_{L^{\frac{4}{3}}} \quad\text{and}\quad \Big\|\int_0^r |f|r'dr'\Big\|_{L^\infty}\lesssim \|f\|_{L^1},$$

where the first inequality is obtained by interpolating the second one and (B.2).

In case of $(i,i')=(1,2)$, we estimate as

$$\begin{aligned} |\mathcal{M}_6| &\lesssim \Big\|\frac{1}{r}\int_0^r \Big|\frac{\psi_1}{r'}\cdot\psi_2\Big|r'dr'\Big\|_{L^4}\Big\|\int_0^r |\psi_3\psi_4|r'dr'\Big\|_{L^\infty}\|\psi_5\psi_6\|_{L^{\frac{4}{3}}} \\ &\lesssim \|r^{-1}\psi_1\|_{L^2}\|\psi_2\|_{L^4}\|\psi_3\|_{L^2}\|\psi_4\|_{L^2}\|\psi_5\|_{L^4}\|\psi_6\|_{L^2}. \end{aligned}$$

In case of $(i,i')\in\{(1,3),(1,5)\}$, we estimate as

$$\begin{aligned} |\mathcal{M}_6| &\lesssim \Big\|\int_0^r \Big|\frac{\psi_1}{r'}\cdot\psi_2\Big|r'dr'\Big\|_{L^\infty}\Big\|\frac{1}{r}\int_0^r |\psi_3\psi_4|r'dr'\Big\|_{L^4}\|\psi_5\psi_6\|_{L^{\frac{4}{3}}} \\ &\lesssim \|r^{-1}\psi_1\|_{L^2}\|\psi_2\|_{L^2}\|\psi_3\|_{L^4}\|\psi_4\|_{L^2}\|\psi_5\|_{L^4}\|\psi_6\|_{L^2}. \end{aligned}$$

In case of $(i,i')\in\{(5,1),(5,6)\}$, we estimate as

$$\begin{aligned} |\mathcal{M}_6| &\lesssim \Big\|\frac{1}{r}\int_0^r |\psi_1\psi_2|r'dr'\Big\|_{L^4}\Big\|\int_0^r |\psi_3\psi_4|r'dr'\Big\|_{L^\infty}\Big\|\frac{\psi_5}{r}\cdot\psi_6\Big\|_{L^{\frac{4}{3}}} \\ &\lesssim \|\psi_1\|_{L^4}\|\psi_2\|_{L^2}\|\psi_3\|_{L^2}\|\psi_4\|_{L^2}\|r^{-1}\psi_5\|_{L^2}\|\psi_6\|_{L^4}. \end{aligned}$$

This completes the proof. □

LEMMA 2.6 ($L^{\frac{4}{3}}$-estimate of $\mathcal{N}$). *Let $m\geq 1$. For any possible choice of $*$ in (2.19) and any index i, there exist $p_j\in\{2,4\}$ $(j\neq i)$ depending on $*$ and i satisfying*

$$\|\mathcal{N}_{3,*}\|_{L^{\frac{4}{3}}}\lesssim \|\psi_i\|_{\dot H^1_m}\prod_{j\neq i}\|\psi_j\|_{L^{p_j}}, \tag{2.33}$$

$$\|\mathcal{N}_{5,*}\|_{L^{\frac{4}{3}}}\lesssim \|\psi_i\|_{\dot H^1_m}\prod_{j\neq i}\|\psi_j\|_{L^{p_j}}, \tag{2.34}$$

and the scaling conditions

$$\sum_{j\neq i} \frac{1}{p_j} = \begin{cases} \frac{3}{4} & \textit{for } (2.33), \\ \frac{7}{4} & \textit{for } (2.34). \end{cases} \tag{2.35}$$

PROOF. The proof follows from Lemma 2.4 and duality relations (2.23). We omit the details. □

2.5. Dynamic rescaling

In works of modulation analysis, one introduces modulation parameters $\lambda(t)$, $\gamma(t)$, and rescale the spacetime variables (t,x) to (s,y). We use $\lambda(t)$ as a scaling parameter function and define the rescaled variables (s,y) by[2]

$$\frac{ds}{dt}=\frac{1}{\lambda^2(t)};\qquad y:=\frac{x}{\lambda(t)}.$$

We work with solutions $u(t,x)$ such that

$$u(t,x)\approx\frac{1}{\lambda(t)}Q\Big(\frac{x}{\lambda(t)}\Big)e^{i\gamma(t)}$$

for some fixed profile Q, and time-dependent scaling and phase parameters $\lambda(t)$ and $\gamma(t)$. Note that u is a function of (t,x) but Q is a function of y. It is thus convenient to introduce notations switching between the (t,x) and (s,y)-variables.

Introduction of $\sharp$ and $\flat$. Let λ and γ be given. When f is a function of y, we use the *raising* operation $\sharp$ to convert f to a function of x. Similarly, for a function $g(x)$, we use the *lowering* operation $\flat$ to convert g to a function of y. We define $\sharp$ and $\flat$ via the formulae[3]

$$f^\sharp(x):=\frac{1}{\lambda}f\Big(\frac{x}{\lambda}\Big)e^{i\gamma},$$
$$g^\flat(y):=\lambda g(\lambda y)e^{-i\gamma}.$$

The rasing/lowering operations respect L^2-scalings in the following sense:

$$\Lambda f^\sharp=[\Lambda f]^\sharp;\qquad if^\sharp=[if]^\sharp;\qquad \|f^\sharp\|_{L^2}=\|f\|_{L^2};$$
$$\Lambda g^\flat=[\Lambda g]^\flat;\qquad ig^\flat=[ig]^\flat;\qquad \|g^\flat\|_{L^2}=\|g\|_{L^2}.$$

Moreover, we have

$$\Delta f^\sharp=\frac{1}{\lambda^2}[\Delta f]^\sharp;\qquad \mathcal{N}(f^\sharp)=\frac{1}{\lambda^2}[\mathcal{N}(f)]^\sharp;\qquad \mathcal{M}(f^\sharp)=\frac{1}{\lambda^2}\mathcal{M}(f);$$
$$\Delta g^\flat=\lambda^2[\Delta g]^\flat;\qquad \mathcal{N}(g^\flat)=\lambda^2[\mathcal{N}(g)]^\flat;\qquad \mathcal{M}(g^\flat)=\lambda^2\mathcal{M}(g).$$

In Chapters 5 and 6, we also use the notation

$$f_b(y):=f(y)e^{-ib\frac{|y|^2}{4}}.$$

Note that $[f_b]^\sharp\neq[f^\sharp]_b$ in general. As we will not use $[f^\sharp]_b$ in any places, we will use the notation

$$f_b^\sharp:=[f_b]^\sharp.$$

[2]Note that $s=s(t)$ is defined up to addition of constants. Only the difference $s(t_1)-s(t_2)$ is of importance.

[3]Musical notations $\sharp$ and $\flat$ are standard in tensor calculus. In our setting, we use $\sharp$ and $\flat$ with completely different meaning; we use them to indicate on which scales we view our functions. But we think that $\sharp$ and $\flat$ in our setting still shares same spirit with those used in tensor calculus.

Introduction of the rescaled variables (s,y)**.** Define the rescaled variables (s,y) by

$$\frac{ds}{dt}=\frac{1}{\lambda^2(t)};\qquad y:=\frac{x}{\lambda(t)}.$$

We will work both on the (t,x) and (s,y)-variables. When we apply $\sharp$ or $\flat$ operations on time-dependent functions, we presume that the time variables t or s are changed in a suitable fashion. For example, $\sharp$ acts on $f(s,y)$ as

$$f^{\sharp}(t,x)=\frac{1}{\lambda(t)}f(s(t),\frac{x}{\lambda(t)})e^{i\gamma(t)}.$$

The following formulae are useful for converting evolutions on (t,x)-variables to those on (s,y)-variables, and vice versa.

$$\begin{aligned}\partial_t f^{\sharp}&=\frac{1}{\lambda^2}\Big[\partial_s f-\frac{\lambda_s}{\lambda}\Lambda f+i\gamma_s f\Big]^{\sharp},\\ \partial_s g^{\flat}&=\lambda^2\Big[\partial_t g+\frac{\lambda_t}{\lambda}\Lambda g-i\gamma_t g\Big]^{\flat}.\end{aligned}$$

For example, if $u(t,x)$ solves (CSS), then we have

$$\begin{aligned}i\partial_t u+\Delta_m u&=\mathcal{N}(u),\\ i\partial_s u^{\flat}+\Delta_m u^{\flat}&=\mathcal{N}(u^{\flat})+i\frac{\lambda_s}{\lambda}\Lambda u^{\flat}+\gamma_s u^{\flat}.\end{aligned}$$

2.6. Local theory of (CSS) under equivariance

The discussion in this section holds for any $g\in\mathbb{R}$ and $m\in\mathbb{Z}$. Under the equivariance, L^2-critical local well-posedness of (1.22) is shown in Liu-Smith [**36**, Section 2] using a standard application of Strichartz estimates. Note that without equivariance ansatz, local theory of (1.1) was developed by varioius authors [**1**, **15**, **35**] (under Coulomb gauge) and [**37**] (under heat gauge). The aim of this section is to record the Strichartz estimates and Cauchy theory of (1.22) under the equivariance. Specializing to the self-dual case $g=1$, the local theory of (CSS) follows.

A pair (q,r) is said to be *admissible* if $2\le q,r\le\infty$, $\frac{1}{q}+\frac{1}{r}=\frac{1}{2}$, and $(q,r)\neq(2,\infty)$. We have the well-known Strichartz estimates:

LEMMA 2.7 (Strichartz estimates). *Let* $\phi:I\times\mathbb{R}^2\to\mathbb{C}$ *solve*

$$\begin{cases}i\partial_t\phi+\Delta\phi=F,\\ \phi(0)=\phi_0,\end{cases}$$

in the sense of Duhamel. For any admissible (q,r) *and* $(\widetilde{q},\widetilde{r})$*, we have*

$$\|\phi\|_{L_I^qL_x^r}\lesssim_{q,r,\widetilde{q},\widetilde{r}}\|\phi_0\|_{L^2}+\|F\|_{L_I^{\widetilde{q}'}L_x^{\widetilde{r}'}},\tag{2.36}$$

where $\widetilde{q}'$ *and* $\widetilde{r}'$ *are conjugate Lebesgue exponents for* $\widetilde{q}$ *and* $\widetilde{r}$*, respectively.*

REMARK 2.8 (Endpoint Strichartz estimate). If we further assume that ϕ is equivariant, then the Strichartz estimates (2.36) hold for the endpoint pairs $(q,r)=(2,\infty)$ or $(\widetilde{q},\widetilde{r})=(2,\infty)$. See the discussion [**36**, Section 2], which is based on [**59**, **60**].

Until the end of this section, $\phi_0^{(n)}$ (with $n \in \mathbb{N}$) and ϕ_0 will always denote initial data. Corresponding maximal lifespan evolution guaranteed by Propositions 2.9 and 2.10 with the initial data $\phi_0^{(n)}$ and ϕ_0 will be denoted by $\phi^{(n)} : I_n \times \mathbb{R}^2 \to \mathbb{C}$ and $\phi : I \times \mathbb{R}^2 \to \mathbb{C}$, respectively.

PROPOSITION 2.9 (L^2-critical Cauchy theory [**36**, Section 2][4]). *Let $m \in \mathbb{Z}$ and $t_0 \in \mathbb{R}$.*

(1) *(Local existence and uniqueness) For any $\phi_0 \in L^2_m$, there exists a unique maximal lifespan solution $\phi : I \times \mathbb{R}^2 \to \mathbb{C}$ in $L^\infty_I L^2_m \cap L^4_{I,\text{loc}} L^4_x$ to (1.22) with the initial data $\phi(t_0) = \phi_0$. Here, I is an open interval containing t_0. Moreover, ϕ indeed lies in $C_{I,\text{loc}} L^2_m$ and $L^q_{I,\text{loc}} L^r_x$ for all admissible pairs (q, r).*
(2) *(Scattering criterion) Solution u scatters forward in time if and only if $\|\phi\|_{L^4_{[t_0,\sup I)} L^4_x} < \infty$. If this is the case, then we have $\sup I = +\infty$. Analogous statement holds for scattering backward in time.*
(3) *(Small data scattering) There exists $\epsilon > 0$ such that if $\|\phi_0\|_{L^2_x} \le \epsilon$, then the solution ϕ satisfies $\|\phi\|_{L^4_{t,x}} < \infty$. In particular, u is global and scatters in both time directions.*
(4) *(Continuous dependence and stability of scattering solutions) Assume that $\{\phi_0^{(n)}\}_{n\in\mathbb{N}} \subset L^2_m$ converges to ϕ_0 strongly in L^2_m. Then, $\liminf_{n\to\infty} I_n \supseteq I$. Moreover, $\phi^{(n)} \to \phi$ in $C_{I,\text{loc}} L^2_m$ and $L^q_{I,\text{loc}} L^r_x$ topology for all admissible pairs (q, r). If furthermore ϕ scatters forward in time in L^2_m, then so does $\phi^{(n)}$ for all large n. Analogous statements holds for scattering backward in time.*

We will also need H^s_m-subcritical local theory for $s > 0$. As the proof is not explicitly given in [**36**], we provide a brief sketch of the proof in Appendix B.

PROPOSITION 2.10 (H^s_m-subcritical Cauchy theory). *Let $m \in \mathbb{Z}$, $t_0 \in \mathbb{R}$, and $s > 0$.*

(1) *(Local existence, uniqueness, and persistence of regularity) For any $\phi_0 \in H^s_m$, there exists a unique maximal lifespan solution $\phi : I \times \mathbb{R}^2 \to \mathbb{C}$ in $L^\infty_{I,\text{loc}} H^s_m \cap L^4_{I,\text{loc}} B^s_{4,2}$ to (1.22) with the initial data $\phi(t_0) = \phi_0$. Here, I is an open interval containing t_0 and equal to that given in Proposition 2.9. Moreover, ϕ indeed lies in $C_{I,\text{loc}} H^s_m$ and $L^q_{I,\text{loc}} B^s_{r,2}$ for any admissible pairs (q, r).*
(2) *(Local existence in subcritical sense) For any $R > 0$, there exists $\delta > 0$ such that we can guarantee $I \supseteq [t_0 - \delta, t_0 + \delta]$ whenever the initial data satisfies $\|\phi_0\|_{H^s_m} \le R$.*
(3) *(Continuous dependence and stability of scattering solutions) The corresponding statement in Proposition 2.9 holds with L^2_m, $C_{I,\text{loc}} L^2_m$, and $L^q_{I,\text{loc}} L^r_x$ replaced by H^s_m, $C_{I,\text{loc}} H^s_m$, and $L^q_{I,\text{loc}} B^s_{r,2}$, respectively.*
(4) *(Finite-time blowup criterion) If $\sup I < +\infty$, then $\|\phi(t)\|_{H^s_m} \to \infty$ as $t \to \sup I$. One can replace $\sup I$ by $\inf I$ to obtain the analogous statement.*

REMARK 2.11. The space $B^s_{r,2}$ is the Besov space, whose definition is given in Appendix B. Such a choice of the solution norm is not important in this paper.

[4]Compared to the statement given in [**36**], we give a qualitative version of continuous dependence. We also include stability of scattering solutions.

CHAPTER 3

Linearization of (CSS) under equivariance

In this chapter, we collect information on the linearization of (CSS). For $g = 1$, the self-dual case, it turns out that the linearized operator at static solution Q is also written as a self-dual form

$$\mathcal{L}_Q = L_Q^* L_Q. \tag{3.1}$$

This is naturally expected in view of the Hamiltonian formulation.

Linearization under the equivariance symmetry was conducted by Lawrie-Oh-Shahshahani [**33**] and they observed that linearized operator has the form (3.1). More precisely, they linearized (CSS) in the form

$$i\partial_t \epsilon - L_Q^* L_Q \epsilon = (\text{h.o.t.}),$$

where $u = Q + \epsilon$ solves (CSS).

In the following, we start with Bogomol'nyi operator and its linearization, and use them to write (CSS) in a self-dual form. Then, we can linearize (CSS) and observe that the linearized operator at Q is of the self-dual form. We study the generalized null space of $i\mathcal{L}_Q$ and coercivity of $\mathcal{L}_Q$.

3.1. Linearization of Bogomol'nyi operator

Because of the Hamiltonian and self-dual structures, it is natural to linearize the Bogomol'nyi operator first. For later use, however, we will linearize at arbitrary profile w. In order to avoid confusion, let us write

$$\begin{aligned} \mathbf{D}_+^{(u)} &:= \partial_r - \tfrac{1}{r}(m + A_\theta[u]), \\ \mathbf{D}_+^{(u)*} &:= -\partial_r - \tfrac{1}{r}(1 + m + A_\theta[u]). \end{aligned}$$

We assume the decomposition

$$u = w + \epsilon.$$

We first observe the operator identities

$$\begin{aligned} \mathbf{D}_+^{(u)} &= \mathbf{D}_+^{(w)} + (B_w \epsilon) + \tfrac{1}{2}(B_\epsilon \epsilon), \\ \mathbf{D}_+^{(u)*} &= \mathbf{D}_+^{(w)*} + (B_w \epsilon) + \tfrac{1}{2}(B_\epsilon \epsilon), \end{aligned}$$

where $(B_w\epsilon)$ (and so is $\frac{1}{2}(B_\epsilon\epsilon)$) is interpreted as multiplication operators by real functions, whose explicit formulae are given by

$$B_f g := \frac{1}{r}\int_0^r \mathrm{Re}(\overline{f} g) r' dr'.$$

We note that

$$A_\theta[u] = -\tfrac{1}{2} r B_u u.$$

We thus have

$$\mathbf{D}_+^{(u)} u = \mathbf{D}_+^{(w)} w + L_w \epsilon + N_w(\epsilon), \tag{3.2}$$

where

$$\begin{aligned} L_w &:= \mathbf{D}_+^{(w)} + w B_w, \\ N_w(\epsilon) &:= \epsilon B_w \epsilon + \tfrac{1}{2} w B_\epsilon \epsilon + \tfrac{1}{2} \epsilon B_\epsilon \epsilon. \end{aligned} \tag{3.3}$$

Here $N_w(\epsilon)$ contains the quadratic and cubic parts of ϵ that are all nonlocal due to integration in r. We remark that the operators B_w and L_w are only $\mathbb{R}$-linear, but *not* $\mathbb{C}$-linear. We also record the formal real adjoints:[1]

$$\begin{aligned} B_w^* f &= w \int_r^\infty (\mathrm{Re} f) dr', \\ L_w^* f &= \mathbf{D}_+^{(w)*} f + B_w^*(\overline{w} f). \end{aligned} \tag{3.4}$$

Note that

$$A_0[u]u = -B_u^*[\tfrac{m}{r}|u|^2 - \tfrac{1}{2}|u|^2 B_u u].$$

In particular when $w = Q$, we use $\mathbf{D}_+^{(Q)} Q = 0$ to have

$$E[Q+\epsilon] = \frac{1}{2} \int |\mathbf{D}_+^{(Q+\epsilon)}(Q+\epsilon)|^2 = \frac{1}{2} \int |L_Q \epsilon + N_Q(\epsilon)|^2.$$

Therefore, we can extract the quadratic part of the energy

$$E[Q+\epsilon] = \frac{1}{2} \int |L_Q \epsilon|^2 + (\text{h.o.t.}). \tag{3.5}$$

3.2. Linearization of (CSS)

In the self-dual case, it is possible to write (CSS) using the (linearized) Bogomol'nyi operator. This is due to the Hamiltonian formulation and a special form of energy (1.29).

PROPOSITION 3.1 ((CSS) in the self-dual form). (CSS) *is equivalent to*

$$i\partial_t u = L_u^* \mathbf{D}_+^{(u)} u. \tag{3.6}$$

REMARK 3.2. As mentioned in Section 1.2, (1.1) under the Coulomb gauge enjoys a Hamiltonian formulation. The equation (CSS) is naturally written as a Hamiltonian equation

$$\partial_t u = -i \frac{\delta E}{\delta u} = -i \frac{\delta}{\delta u} \Big(\frac{1}{2} \int |\mathbf{D}_+^{(u)} u|^2 \Big).$$

Using the linearization (3.2), one can formally verify (3.6).

PROOF OF PROPOSITION 3.1. Recall that

$$\begin{aligned} \mathbf{D}_+^{(u)} u &= (\partial_r - \tfrac{m}{r}) u + \tfrac{1}{2}(B_u u) u, \\ L_u^* &= (-\partial_r - \tfrac{m+1}{r}) + \tfrac{1}{2}(B_u u) + B_u^*[\overline{u} \cdot]. \end{aligned}$$

[1] By real adjoint, we view $L^2(\mathbb{R}^2; \mathbb{C})$ as $\mathbb{R}$-Hilbert space equipped with the inner product $(u,v)_r = \int \mathrm{Re}(u\overline{v})$.

Thus

$$\begin{aligned}L_u^*\mathbf{D}_+^{(u)}u &= (-\partial_r - \tfrac{m+1}{r})(\partial_r - \tfrac{m}{r})u \\ &\quad + (-\partial_r - \tfrac{m+1}{r})[\tfrac{1}{2}(B_u u)u] + \tfrac{1}{2}(B_u u)(\partial_r - \tfrac{m}{r})u + B_u^*[\overline{u}(\partial_r - \tfrac{m}{r})u] \\ &\quad + \tfrac{1}{4}[B_u u]^2 u + \tfrac{1}{2}B_u^*[|u|^2 B_u u].\end{aligned}$$

We can rewrite the terms in the second line of the above display as

$$\begin{aligned}&(-\partial_r - \tfrac{m+1}{r})[\tfrac{1}{2}(B_u u)u] + \tfrac{1}{2}(B_u u)(\partial_r - \tfrac{m}{r})u \\ &= \tfrac{1}{2r}(-\partial_r - \tfrac{m}{r})[r(B_u u)u] + \tfrac{1}{2r}(rB_u u)(\partial_r - \tfrac{m}{r})u \\ &= \tfrac{1}{2r}[-\partial_r, r(B_u u)]u - \tfrac{m}{r}(B_u u)u \\ &= -\tfrac{1}{2}|u|^2 u - \tfrac{m}{r}(B_u u)u,\end{aligned}$$

and

$$B_u^*[\overline{u}(\partial_r - \tfrac{m}{r})u] = -\tfrac{1}{2}|u|^2 u - B_u^*[\tfrac{m}{r}|u|^2].$$

Therefore, we recover the nonlinearity expression as in (CSS), but expressed in terms of B_u, B_u^*:

$$\begin{aligned}L_u^*\mathbf{D}_+^{(u)}u &= (-\partial_r - \tfrac{m+1}{r})(\partial_r - \tfrac{m}{r})u \\ &\quad - |u|^2 u - \tfrac{m}{r}(B_u u)u - B_u^*[\tfrac{m}{r}|u|^2] \\ &\quad + \tfrac{1}{4}(B_u u)^2 u + \tfrac{1}{2}B_u^*[|u|^2 B_u u] \\ &= -\Delta_m u - |u|^2 u + \frac{2m}{r^2}A_\theta[u]u + \frac{A_\theta^2[u]}{r^2}u + A_0[u]u.\end{aligned}$$

This completes the proof. □

Using (3.6), we can easily linearize (CSS). To linearize $L_u^*\mathbf{D}_+^{(u)}u$ at w, we write

$$\begin{aligned}u &= w + \epsilon, \\ \mathbf{D}_+^{(u)}u &= \mathbf{D}_+^{(w)}w + L_w\epsilon + N_w(\epsilon), \\ L_u^* &= L_w^* + [(B_w\epsilon) + B_w^*[\overline{\epsilon}\,\cdot\,] + B_\epsilon^*[\overline{w}\,\cdot\,]] + [\tfrac{1}{2}(B_\epsilon\epsilon) + B_\epsilon^*[\overline{\epsilon}\,\cdot\,]].\end{aligned}$$

Then the linear part $\mathcal{L}_w\epsilon$ of $L_u^*\mathbf{D}_+^{(u)}u$ is given by

$$\mathcal{L}_w\epsilon = L_w^*L_w\epsilon + [(B_w\epsilon) + B_w^*[\overline{\epsilon}\,\cdot\,] + B_\epsilon^*[\overline{w}\,\cdot\,]]\mathbf{D}_+^{(w)}w. \tag{3.7}$$

In particular, when $w = Q$, we observe the *self-duality* using $\mathbf{D}_+^{(Q)}Q = 0$:

$$\mathcal{L}_Q = L_Q^*L_Q. \tag{3.8}$$

With this formulation, we can recover a full linearized equation

$$\begin{aligned}i\partial_t\epsilon - \mathcal{L}_Q\epsilon &= L_Q^*N_Q(\epsilon) + [(B_Q\epsilon) + B_Q^*[\overline{\epsilon}\,\cdot\,] + B_\epsilon^*[Q\,\cdot\,]][L_Q\epsilon + N_Q(\epsilon)] \\ &\quad + [\tfrac{1}{2}(B_\epsilon\epsilon) + B_\epsilon^*[\overline{\epsilon}\,\cdot\,]][L_Q\epsilon + N_Q(\epsilon)],\end{aligned}$$

as is done in Lawrie-Oh-Shahshahani [**33**]. They also observed that there is no derivative falling on ϵ in the nonlinearity. This is naturally expected as the nonlinearity of (CSS) contains no derivatives.

3.3. Algebraic relations, solvability, and coercivity of $\mathcal{L}_Q$

The linearization of (CSS) at Q is given by

$$\partial_t \epsilon = -i\mathcal{L}_Q \epsilon.$$

Our main goals are to compute the generalized nullspace of $i\mathcal{L}_Q$ and to prove the coercivity of $\mathcal{L}_Q$ under a transversality condition. Such spectral information stems from the symmetries of (CSS), i.e. phase, scaling, and pseudoconformal symmetries. It turns out that the generalized null space for the *self-dual* (CSS) is different from that of (NLS) studied in [**62**].

Differentiating symmetries of (CSS) at the static solution Q, we obtain explicit algebraic identities satisfied by $\mathcal{L}_Q$. Assume we have a continuous family of solutions $u^{(a)}(t,r)$ to (CSS) with $u^{(0)}(t,r) = Q(r)$. Differentiating

$$L_{u^{(a)}}^* \mathbf{D}_+^{(u^{(a)})} u^{(a)} = i\partial_t u^{(a)}$$

in a at $a = 0$, we obtain

$$\mathcal{L}_Q(\partial_a u^{(a)})|_{a=0} = i\partial_t(\partial_a u^{(a)})|_{a=0}.$$

If we substitute phase, scaling, and pseudoconformal symmetry

$$u^{(a)}(t,r) = \begin{cases} e^{ia}Q(r) & \text{(phase)} \\ aQ(ar) & \text{(scaling)} \\ \frac{1}{1+at} e^{ia\frac{r^2}{4(1+at)}} Q(\frac{r}{1+at}) & \text{(pseudoconformal)} \end{cases}$$

then we obtain

$$\begin{aligned} \mathcal{L}_Q(iQ) &= 0, && \text{(phase)} \\ \mathcal{L}_Q(\Lambda Q) &= 0, && \text{(scaling)} \\ \mathcal{L}_Q(ir^2 Q) &= -4i\Lambda Q. && \text{(pseudoconformal)} \end{aligned}$$

Here is a rough discussion on spectral properties of $\mathcal{L}_Q$. Rigorous results are proved below. Noticing that $\mathcal{L}_Q f = 0$ if and only if $L_Q f = 0$, we have $L_Q(iQ) = L_Q(\Lambda Q) = 0$. As L_Q is a first-order (nonlocal) differential operator, one can indeed see that $\ker L_Q = \mathrm{span}_{\mathbb{R}}\{iQ, \Lambda Q\}$. Moreover, one can invert the operator $\mathcal{L}_Q$ on the orthogonal complement of $\ker L_Q$. As a consequence, $\mathcal{L}_Q$ is coercive on a subspace having trivial intersection with $\ker L_Q$.

As an application, we can characterize the generalized null space of $i\mathcal{L}_Q$. We first show that the kernel of L_Q (and hence of $\mathcal{L}_Q$) is $\mathrm{span}_{\mathbb{R}}\{\Lambda Q, iQ\}$.

LEMMA 3.3 (Kernel of L_Q). *If $f : \mathbb{R}^2 \to \mathbb{C}$ is a smooth m-equivariant function such that $L_Q f = 0$,[2] then f is a $\mathbb{R}$-linear combination of ΛQ and iQ.*

PROOF. As f is smooth m-equivariant, Lemma A.3 says that $f^{(k)}(0) = 0$ for all $k \le m-1$. As $\varphi \in \{\Lambda Q, iQ\}$ satisfies $\varphi^{(k)}(0) = 0$ for $k \le m-1$ and $\varphi^{(m)}(0) \neq 0$, we may subtract an $\mathbb{R}$-linear combination of ΛQ and iQ from f to assume $f^{(m)}(0) = 0$. It now suffices to show that $f = 0$. Since f is smooth such that $f^{(k)}(0) = 0$ for all $k \le m$, by Lemma A.1, the function $r^{-m}f(r)$ is smooth, $r^{-m}f(r)|_{r=0} = 0$, and

$$\partial_r(r^{-m}f) = r^{-m}(\partial_r f - \tfrac{m}{r}f) = -r^{-m}(QB_Q + \tfrac{1}{2}(B_Q Q))f.$$

[2] Actually L_Q acts on functions from $(0,\infty)$ to $\mathbb{C}$. By an abuse of notation, we denote by $f(r)$ the radial component of the m-equivariant function $f(x)$.

We integrate from the origin to get

$$\begin{aligned}\frac{|f(r)|}{r^m} &= \Big|\int_0^r \frac{Q}{(r')^{m+1}}\Big(\int_0^{r'} Q\mathrm{Re} f dr''\Big)dr' + \frac{1}{2}\int_0^r (B_Q Q)\frac{f(r')}{(r')^m}dr'\Big| \\ &\le \int_0^r \Big\{(r')^{m+1}Q\Big(\int_{r'}^r \frac{Q}{(r'')^{m+1}}dr''\Big) + \frac{1}{2}B_Q Q\Big\}\frac{|f(r')|}{(r')^m}dr'\end{aligned}$$

As the factor in the curly bracket is bounded, Gronwall's inequality concludes that $f = 0$. □

To find the generalized null space of $i\mathcal{L}_Q$, it is natural to ask whether we can solve the equations $i\mathcal{L}_Q\psi = \Lambda Q$ and $i\mathcal{L}_Q\varphi = iQ$. More generally, we may consider the equation $i\mathcal{L}_Q f = h$, where h is given. We will check solvability in the following sense. If there is a nice function[3] f with $i\mathcal{L}_Q f = h$, then

$$(h, ig)_r = (i\mathcal{L}_Q f, ig)_r = (f, \mathcal{L}_Q g)_r \tag{3.9}$$

must vanish for any $g \in \{\Lambda Q, iQ\}$. Thus if $(h, i\Lambda Q)_r$ or $(h, Q)_r$ does not vanish, then we say that $i\mathcal{L}_Q f = h$ is *not solvable*. On the other hand, if $(h, i\Lambda Q)_r = (h, Q)_r = 0$, then we are able to find f (thus *solvable*) for sufficiently nice h as in the proof of Lemma 3.6 below.

It is natural to expect that we can solve $i\mathcal{L}_Q\psi = \Lambda Q$ and $i\mathcal{L}_Q\psi = iQ$ due to $(\Lambda Q, Q)_r = 0$. For the former, we already know a solution, which is $i\frac{1}{4}r^2 Q$ given by the pseudoconformal symmetry. But there is no ψ satisfying $i\mathcal{L}_Q\psi = i\frac{1}{4}r^2Q$ because $(r^2Q, \Lambda Q)_r = -\|rQ\|_{L^2}^2 \neq 0$ (when $m \ge 1$). For the latter, we can in fact construct a solution ρ to $i\mathcal{L}_Q\rho = iQ$ in Lemma 3.6 below. But again there is no ψ satisfying $i\mathcal{L}_Q\psi = \rho$ because $(-i\rho, iQ)_r = -\|L_Q\rho\|_{L^2}^2 \neq 0$ (when $m \ge 1$). Therefore, we have found the basis of the generalized null space. In summary,

PROPOSITION 3.4 (The generalized null space of $i\mathcal{L}_Q$). *The generalized null space*[4] *of $i\mathcal{L}_Q$ has $\mathbb{R}$-basis $\{iQ, \Lambda Q, ir^2Q, \rho\}$ with relations*

$$\begin{aligned} i\mathcal{L}_Q\rho &= iQ; & i\mathcal{L}_Q ir^2Q &= 4\Lambda Q;\\ i\mathcal{L}_Q iQ &= 0; & i\mathcal{L}_Q\Lambda Q &= 0;\end{aligned}$$

where ρ is given in Lemma 3.6 below.

REMARK 3.5. It is instructive to compare with (NLS). Recall the ground state R (1.34) and associated linearized operator $\mathcal{L}_{\mathrm{NLS}}$ (1.36)

$$\mathcal{L}_{\mathrm{NLS}} f = -\Delta f + f - 2R^2 f - R^2\overline{f}.$$

It is known from [**62**] that under radial symmetry, the generalized null space of $i\mathcal{L}_{\mathrm{NLS}}$ is characterized as

$$\begin{aligned} i\mathcal{L}_{\mathrm{NLS}}\rho_{\mathrm{NLS}} &= ir^2R,\\ i\mathcal{L}_{\mathrm{NLS}} i|y|^2R &= 4\Lambda R,\\ i\mathcal{L}_{\mathrm{NLS}}\Lambda R &= -2iR,\\ i\mathcal{L}_{\mathrm{NLS}} iR &= 0.\end{aligned}$$

[3]We will not be specific here, a nice function is required to make sense of (3.9). One may require smoothness, integrability, or boundary conditions (at $r = 0$ or $r = \infty$) for such functions.

[4]In fact, when $m = 0$ or 1, ir^2Q and ρ do not belong to L^2 . We will still call them generalized eigenmodes due to the algebraic identities in this proposition.

Notice that ΛR does not belong to the kernel of $i\mathcal{L}_{\mathrm{NLS}}$. This is because $e^{it}R(x)$ is not a static solution to (NLS). Moreover, $\mathcal{L}_{\mathrm{NLS}}$ restricted on real-valued radial functions is invertible. Thus there exists real-valued ρ_{NLS} solving $\mathcal{L}_{\mathrm{NLS}}\rho_{\mathrm{NLS}} = r^2 R$. This is in strong contrast to the case of (CSS).

LEMMA 3.6 (The generalized eigenmode ρ). *There exists a smooth real-valued solution $\rho : (0,\infty) \to \mathbb{R}$ to*

$$L_Q^* L_Q \rho = Q$$

satisfying

$$|\rho(r)| \lesssim r^2 Q \sim \begin{cases} r^{m+2} & \text{if } r \le 1, \\ r^{-m} & \text{if } r > 1. \end{cases}$$

Moreover, if $m \geq 1$, we have the nondegeneracy[5]

$$(\rho, Q)_r = \|L_Q \rho\|_{L^2}^2 \neq 0.$$

PROOF. In the proof, we present how we can solve $L_Q^* L_Q \rho = \varphi$ whenever φ is a generic real-valued function with $(\varphi, \Lambda Q)_r = 0$. For sake of simplicity, however, we only present the special case when $\varphi = Q$.

At first, we construct a real-valued function ψ satisfying

$$L_Q^* \psi = Q; \qquad |\psi(r)| \lesssim rQ \sim \begin{cases} r^{m+1} & \text{if } r \le 1, \\ r^{-m-1} & \text{if } r > 1. \end{cases} \tag{3.10}$$

Rewrite $L_Q^* \psi = Q$ as

$$-\partial_r \psi - \frac{1}{r}(1 + m + A_\theta[Q])\psi + Q \int_r^\infty Q\psi dr' = Q.$$

Using the relation $\partial_r Q - \frac{1}{r}(m + A_\theta[Q])Q = \mathbf{D}_+^{(Q)} Q = 0$, we renormalize the equation using $\widetilde{\psi} := (rQ)\psi$ as

$$-\partial_r \widetilde{\psi} + rQ^2 \int_r^\infty \widetilde{\psi} \frac{dr'}{r'} = rQ^2.$$

We integrate from the spatial infinity with $\widetilde{\psi}(\infty) = 0$ to get the integral equation

$$\widetilde{\psi}(r) + \int_r^\infty Q^2 \Big(\int_{r'}^\infty \widetilde{\psi} \frac{dr''}{r''} \Big) r' dr' = \int_r^\infty Q^2 r' dr'. \tag{3.11}$$

As $Q^2 r \lesssim r^{-2m-3}$ for $r \gg 1$, we seek for $\widetilde{\psi}(r)$ having decay r^{-2m-2}. Note that

$$\begin{aligned} \int_r^\infty Q^2 \Big(\int_{r'}^\infty r^{-2m-2} \frac{dr''}{r''} \Big) r' dr' &\le \frac{1}{2m+2} \int_r^\infty Q^2 (r')^{-2m-1} dr' \\ &\le \frac{\|rQ\|_{L^\infty}^2}{(2m+2)^2} r^{-2m-2} = \frac{1}{2} r^{-2m-2}. \end{aligned}$$

As the coefficient of $\frac{1}{2} r^{-2m-2}$ is less than 1, a standard contraction principle allows us to construct $\widetilde{\psi}$ satisfying

$$\widetilde{\psi}(r) \lesssim r^{-2m-2}, \qquad \forall r > 0.$$

In other words, we have

$$\psi(r) \lesssim (rQ)^{-1} r^{-2m-2}, \qquad \forall r > 0,$$

which shows r^{-m-1} decay at infinity.

[5] If $m = 0$, then ρ has tail ~ 1 as $r \to \infty$, so $(\rho, Q)_r$ is not defined.

In order to show the decay of ψ at the origin, we use a recursive argument. We crucially exploit the facts $(Q, \Lambda Q)_r = 0$ and $L_Q \Lambda Q = 0$. For any $0 < \delta \leq 1$, we observe that

$$|(L_Q^* \psi, \mathbf{1}_{r \geq \delta} \Lambda Q)| = |(Q, \mathbf{1}_{r \geq \delta} \Lambda Q)| = |(Q, \mathbf{1}_{r < \delta} \Lambda Q)| \lesssim \delta^{2m+2}.$$

On the other hand, we have

$$\begin{aligned}
&(L_Q^* \psi, \mathbf{1}_{r \geq \delta} \Lambda Q) \\
&= \psi(\delta) \Lambda Q(\delta) \delta + \int_\delta^\infty \psi\Big((\partial_r - \frac{m + A_\theta[Q]}{r}) \Lambda Q + \frac{Q}{r} \int_\delta^r Q \Lambda Q r' dr' \Big) r dr \\
&= \psi(\delta) \Lambda Q(\delta) \delta - \int_\delta^\infty \psi Q \Big(\int_0^\delta Q \Lambda Q r' dr' \Big) dr.
\end{aligned}$$

Note that

$$\Lambda Q(\delta) \delta \sim \delta^{m+1},$$

$$\Big| \int_\delta^\infty \psi Q \Big(\int_0^\delta Q \Lambda Q r' dr' \Big) dr \Big| \lesssim \delta^{2m+2} \int_\delta^\infty |\widetilde{\psi}(r)| \frac{dr}{r}.$$

Recalling $\widetilde{\psi} = rQ\psi$, we arrange the above estimate to obtain the recursive estimate

$$|\psi(\delta)| \lesssim \delta^{m+1} \Big(1 + \int_\delta^\infty rQ|\psi| \frac{dr}{r} \Big). \tag{3.12}$$

Starting from our rough bound $\psi(r) \lesssim (rQ)^{-1} r^{-2m-2}$, iterating (3.12) three times yields $|\psi(\delta)| \lesssim \delta^{m+1}$, which is the desired behavior near the origin. This completes the proof of (3.10).

Next, we construct a real-valued function ρ satisfying

$$L_Q \rho = \psi, \qquad |\rho(r)| \lesssim r^2 Q \sim \begin{cases} r^{m+2} & \text{if } r \leq 1, \\ r^{-m} & \text{if } r \geq 1. \end{cases} \tag{3.13}$$

We solve this by integrating from the origin. Rewrite $L_Q \rho = \psi$ as

$$\partial_r \rho - \frac{1}{r}(m + A_\theta[Q]) \rho + \frac{Q}{r} \int_0^r Q \rho r' dr' = \psi.$$

We conjugate Q to ρ, i.e. we consider $\rho = Q\widetilde{\rho}$ and rewrite the equation as

$$\partial_r \widetilde{\rho} + \frac{1}{r} \int_0^r Q^2 \widetilde{\rho} r' dr' = Q^{-1} \psi.$$

We integrate from the origin with $\widetilde{\rho}(0) = 0$ to write

$$\widetilde{\rho}(r) + \int_0^r \Big(\int_0^{r'} Q^2 \widetilde{\rho} r'' dr'' \Big) \frac{dr'}{r'} = \int_0^r Q^{-1} \psi dr'. \tag{3.14}$$

As $Q^{-1}\psi \lesssim r$ near the origin, we expect $\widetilde{\rho}(r) \lesssim r^2$ near the origin. Since

$$\int_0^r \Big(\int_0^{r'} Q^2 (r'')^2 r'' dr'' \Big) \frac{dr'}{r'} \lesssim r^{2m+2} \ll r^2, \qquad \forall r \ll 1,$$

a standard contraction principle allows us to construct $\widetilde{\rho}$ on $[0, r_0)$ for some $r_0 \ll 1$ with expected pointwise bound $\widetilde{\rho}(r) \lesssim r^2$.

To construct $\widetilde{\rho}$ beyond $[0, r_0)$, choose an increasing sequence $\{r_n\}_{n \geq 0} \subset \mathbb{R}$ such that $r_n \to \infty$, $r_{n+1} \leq 2r_n$, and $\int_{r_n}^{r_{n+1}} Q^2 r'' dr'' \ll 1$. We will construct $\widetilde{\rho}$ on

$[r_n, r_{n+1})$ inductively on n. Suppose we have constructed $\widetilde{\rho}$ on the interval $[0, r_n)$ for some $n \geq 0$. We then rewrite the integral equation of $\widetilde{\rho}$ as

$$\widetilde{\rho}(r) + \int_{r_n}^{r} \Big(\int_{r_n}^{r'} Q^2 \widetilde{\rho} r'' dr'' \Big) \frac{dr'}{r'} = \int_0^r \Big(Q^{-1}\psi - \frac{1}{r'} \int_0^{\min(r_n, r')} Q^2 \widetilde{\rho} r'' dr'' \Big) dr'$$

for $r \in [r_n, r_{n+1})$. Since (by Fubini and $r_{n+1} \leq 2r_n$)

$$\int_{r_n}^{r} \Big(\int_{r_n}^{r'} Q^2 r'' dr'' \Big) \frac{dr'}{r'} \leq 2 \int_{r_n}^{r_{n+1}} Q^2 r'' dr'' \ll 1,$$

we can use contraction principle to construct $\widetilde{\rho}(r)$ on $[r_n, r_{n+1})$.

From the above recursive construction, we have defined $\widetilde{\rho}$ globally on $[0, +\infty)$ but the growth of $\widetilde{\rho}$ can be wild. In order to obtain the desired growth $\widetilde{\rho}(r) \lesssim r^2$, choose a positive increasing sequence $\{C_n\}_{n \in \mathbb{N}} \subset \mathbb{R}$ such that $|\widetilde{\rho}(r)| \leq C_n r^2$ for all $r \in [0, r_n]$ and $n \in \mathbb{N}$. We claim that there exist large $N \in \mathbb{N}$ and $C > C_N$ such that $|\widetilde{\rho}(r)| \leq Cr^2$ for all $r \geq r_N$. To see this, substitute the bootstrap hypothesis $|\widetilde{\rho}(r)| \leq Cr^2$ for $r \geq r_N$ into (3.14) to get

$$|\widetilde{\rho}(r)| \lesssim r^2 + C \log^2 r$$

for $r \geq r_N$. If we choose N large such that $\log^2 r \ll r^2$ for $r \geq r_N$ and $C > C_N$ large, we get a strong conclusion $|\widetilde{\rho}(r)| \leq \frac{1}{2} Cr^2$. Therefore, we obtain $\widetilde{\rho}(r) \leq Cr^2$ by the continuity argument. This translates to the desired bound

$$\rho(r) \lesssim r^2 Q.$$

This completes the proof of (3.13).

In case of $m \geq 1$, note that $Q\rho$ is integrable. Thus we get the nondegeneracy

$$(Q, \rho)_r = (L_Q^* L_Q \rho, \rho)_r = \|L_Q \rho\|_{L^2}^2 \neq 0.$$

This completes the proof. □

REMARK 3.7 (Explicit construction of ψ). One can find the explicit formula of ψ to (3.10) as

$$L_Q \rho = \psi = \frac{1}{2(m+1)} rQ = \sqrt{2} \frac{r^{m+1}}{1 + r^{2m+2}}. \tag{3.15}$$

One can easily verify that this ψ satisfies $L_Q^* \psi = Q$, though we found this formula by writing an ansatz (recall $\widetilde{\psi} = rQ\psi$)

$$\widetilde{\psi} = \sum_{k=1}^{\infty} \frac{c_k}{(1 + r^{2m+2})^k}$$

with $c_1 = 4(m+1)$. Note that this ansatz is closed under the iteration scheme (3.11); we have

$$\begin{aligned} &\int_r^\infty Q^2 \Big(\int_{r'}^\infty \frac{1}{(1 + (r'')^{2m+2})^k} \frac{dr''}{r''} \Big) r' dr' \\ &= \int_r^\infty Q^2 \Big(\sum_{\ell \geq k} \frac{1}{(2m+2)\ell} \cdot \frac{1}{(1 + (r')^{2m+2})^\ell} \Big) r' dr' \\ &= \sum_{\ell \geq k} \frac{2}{\ell(\ell+1)} \cdot \frac{1}{(1 + r^{2m+2})^{\ell+1}}. \end{aligned}$$

Starting from $c_1 = 4(m+1)$, the equation (3.11) determines all the other coefficients $c_2 = -4(m+1)$ and $c_3 = c_4 = \cdots = 0$. This shows

$$\widetilde{\psi} = 4(m+1)\frac{r^{2m+2}}{(1+r^{2m+2})^2}$$

and hence (3.15).

REMARK 3.8 (Asymptotics of ρ). For $m \geq 1$, we claim that

$$\rho(r) = \Big(\frac{1}{4(m+1)}r^2 - \frac{\|\psi\|_{L^2}^2}{2\pi}\log r + O(1)\Big)Q(r) \tag{3.16}$$

as $r \to \infty$. Indeed, the relation $L_Q^* L_Q \rho = L_Q^* \psi = Q$ and estimate $\widetilde{\rho} \lesssim r^2$ (recall $\widetilde{\rho} = Q^{-1}\rho$) say that

$$\int_0^{r'} Q^2 \widetilde{\rho} r'' dr'' = \int_0^\infty Q\rho r'' dr'' - \int_{r'}^\infty Q^2 \widetilde{\rho} r'' dr'' = \frac{\|\psi\|_{L^2}^2}{2\pi} - O((r')^{-2m})$$

as $r \to \infty$. Substituting this and (3.15) into the integral equation (3.14) shows

$$\widetilde{\rho}(r) = \frac{1}{4(m+1)}r^2 - \frac{\|\psi\|_{L^2}^2}{2\pi}\log r + O(1)$$

as $r \to \infty$. This completes the proof of (3.16).

We conclude this section with the coercivity estimate. Thanks to the factorization $\mathcal{L}_Q = L_Q^* L_Q$, coercivity of $\mathcal{L}_Q$ easily follows from the characterization of the kernel of L_Q.

LEMMA 3.9 (Coercivity of L_Q). *Let $m \geq 1$. We have*

$$\|L_Q f\|_{L^2} \lesssim \|f\|_{\dot{H}_m^1}, \qquad \forall f \in \dot{H}_m^1 \tag{3.17}$$

$$\|L_Q f\|_{L^2}^2 \geq c_{\psi_1,\psi_2}\|f\|_{\dot{H}_m^1}^2, \qquad \forall f \in \dot{H}_m^1,\ f \perp \{\psi_1, \psi_2\}, \tag{3.18}$$

where $\{\psi_1, \psi_2\} \subset (\dot{H}_m^1)^*$ are such that the following matrix has nonzero determinant:

$$\begin{bmatrix} (iQ,\psi_1)_r & (\Lambda Q,\psi_1)_r \\ (iQ,\psi_2)_r & (\Lambda Q,\psi_2)_r \end{bmatrix}$$

In case of $m = 0$, (3.17) and (3.18) hold true if one replaces $\dot{H}_m^1$ by

$$\|f\|_{\dot{\mathcal{H}}_0}^2 := \|\partial_r f\|_{L^2}^2 + \|(1+r)^{-1} f\|_{L^2}^2.$$

PROOF. For simplicity of exposition, we hope to introduce a unified notation

$$\|f\|_{\dot{\mathcal{H}}_m}^2 := \begin{cases} \|\partial_r f\|_{L^2}^2 + \|(1+r)^{-1} f\|_{L^2}^2 & \text{if } m = 0, \\ \|f\|_{\dot{H}_m^1}^2 & \text{if } m \geq 1, \end{cases}$$

We first show (3.17). Since

$$\|L_Q f\|_{L^2} \lesssim \|f\|_{\dot{\mathcal{H}}_m} + \|QB_Q f\|_{L^2},$$

it suffices to control the term $QB_Q f$. For this purpose, we express

$$[QB_Q f](r) = \int_0^\infty K(r,r')\frac{\mathrm{Re} f(r')}{1+r'} r' dr',$$

$$K(r,r') := r^{-1}Q(r)Q(r')(1+r')\mathbf{1}_{r' \leq r}.$$

Since

$$\sup_{r>0}\int_0^\infty |K(r,r')|r'dr' \lesssim \|Q\|_{L^\infty}\sup_{r>0}\Big(\frac{1}{r}\int_0^r (1+r')Q(r')r'dr'\Big) \lesssim 1,$$
$$\sup_{r'>0}\int_0^\infty |K(r,r')|rdr \lesssim \|Q\|_{L^\infty}\sup_{r'>0}\Big((1+r')\int_{r'}^\infty Q(r)dr\Big) \lesssim 1,$$

Schur's test yields

$$\|QB_Qf\|_{L^2} \lesssim \|(1+r)^{-1}f\|_{L^2} \lesssim \|f\|_{\dot{\mathcal{H}}_m}.$$

This completes the proof of (3.17).

We turn to (3.18). We assert the following weak coercivity:

$$\|L_Qf\|_{L^2} \geq c_{\psi_1,\psi_2}\begin{cases}\|(1+r)^{-1}f\|_{L^2} & \text{if } m=0,\\ \|r^{-1}f\|_{L^2} & \text{if } m\geq 1,\end{cases} \qquad \forall f\perp\{\psi_1,\psi_2\}. \tag{3.19}$$

We follow the standard strategy used in the proof of Poincaré inequality. If (3.19) fails, then there exists a sequence $\{f_n\}_{n\in\mathbb{N}}\perp\{\psi_1,\psi_2\}$ such that

$$\|L_Qf_n\|_{L^2}\to 0 \quad\text{and}\quad \begin{cases}\|(1+r)^{-1}f_n\|_{L^2}=1 & \text{if } m=0,\\ \|r^{-1}f_n\|_{L^2}=1 & \text{if } m\geq 1.\end{cases}$$

We claim for such a sequence that there exist $0<r_1\ll 1$ and $r_2\gg 1$ with

$$\begin{cases}\liminf_{n\to\infty}\|f_n\|_{L^2(|x|<r_2)}>0, & \text{if } m=0,\\ \liminf_{n\to\infty}\|f_n\|_{L^2(r_1<|x|<r_2)}>0, & \text{if } m\geq 1.\end{cases} \tag{3.20}$$

Let us assume this claim and finish the proof of (3.19). As all the terms of L_Qf except ∂_rf can be controlled by $(1+r)^{-1}f$ if $m=0$ and $r^{-1}f$ if $m\geq 1$, we see that $\|\partial_rf_n\|_{L^2}$ is bounded. Passing to a subsequence, we can assume $f_n\rightharpoonup f$ in $\dot{\mathcal{H}}_m$. Moreover, we can apply the Rellich-Kondrachov compactness lemma (on the region $\{|x|<r_2\}$ if $m=0$ and $\{r_1<|x|<r_2\}$ if $m\geq 1$) to guarantee that $f\neq 0$. From the weak convergence, we have $L_Qf=0$ (weakly) and $(f,\psi_1)_r=(f,\psi_2)_r=0$. We understand $L_Qf=0$ on the ambient space $\mathbb{R}^2$ as

$$(\partial_1+i\partial_2)[f(r)e^{im\theta}] = \Big(\frac{A_\theta[Q]}{r}f-QB_Qf\Big)(r)e^{i(m+1)\theta}.$$

By elliptic regularity, f is a smooth m-equivariant solution. By Lemma 3.3, f is an $\mathbb{R}$-linear combination of ΛQ and iQ. Because of orthogonality conditions, we should have $f=0$, yielding a contradiction. This completes the proof of the weak coercivity (3.19) assuming the claim (3.20).

We turn to show the claim (3.20). Suppose not; we can choose a subsequence (still denoted by $\{f_n\}_{n\in\mathbb{N}}$) such that

$$\begin{cases}\lim_{n\to\infty}\|f_n\|_{L^2(|x|<r_2)}=0, & \text{if } m=0,\\ \lim_{n\to\infty}\|f_n\|_{L^2(r_1<|x|<r_2)}=0, & \text{if } m\geq 1,\end{cases}$$

for any $0 < r_1 < r_2 < \infty$. We first observe the following computation: for any $0 < y_1 < y_2 < \infty$,

$$\begin{aligned}\int_{y_1}^{y_2} \Big|L_Q f_n - \frac{f_n}{r}\Big|^2 r dr &= \int_{y_1}^{y_2} |L_Q f_n|^2 r dr - 2\int_{y_1}^{y_2} \mathrm{Re}(\overline{f}_n L_Q f_n) dr + \int_{y_1}^{y_2} \Big|\frac{f_n}{r}\Big|^2 r dr \\ &= \int_{y_1}^{y_2} |L_Q f_n|^2 r dr - |f_n|^2\Big|_{y_1}^{y_2} - 2\int_{y_1}^{y_2} \mathrm{Re}\Big(\frac{f_n}{r}\Big) Q B_Q(f_n) r dr \\ &\quad + \int_{y_1}^{y_2} (1 + 2(m + A_\theta[Q]))\Big|\frac{f_n}{r}\Big|^2 r dr.\end{aligned}$$

Since $A_\theta[Q](r) \to -2(m+1)$ as $r \to \infty$, we take $y_1 = r_2$ for sufficiently large r_2 and $y_2 \to \infty$ to obtain

$$\int_{r_2}^{\infty} \Big|\frac{f_n}{r}\Big|^2 r dr \le |f_n|^2(r_2) + \int_{r_2}^{\infty} |L_Q f_n|^2 r dr + 2\int_{r_2}^{\infty} \Big|\frac{f_n}{r}\Big| Q B_Q(|f_n|) r dr.$$

The last term of the RHS can be absorbed into the LHS as r_2 is large. We then average in the r_2-variable and take limit $n \to \infty$ to get

$$\liminf_{n\to\infty} \int_{2r_2}^{\infty} \Big|\frac{f_n}{r}\Big|^2 r dr \lesssim \liminf_{n\to\infty} \int_{r_2}^{2r_2} \Big|\frac{f_n}{r}\Big|^2 r dr. \tag{3.21}$$

If $m = 0$, then we get a contradiction. Hence (3.20) is true for $m = 0$.

When $m \ge 1$, we should prevent f_n from concentrating at the origin. For this, we need one more computation: for any $0 < y_1 < y_2 < \infty$,

$$\begin{aligned}\int_{y_1}^{y_2} \Big|L_Q f_n + \frac{f_n}{r}\Big|^2 r dr &= \int_{y_1}^{y_2} |L_Q f_n|^2 r dr + |f_n|^2\Big|_{y_1}^{y_2} + 2\int_{y_1}^{y_2} \mathrm{Re}\Big(\frac{f_n}{r}\Big) Q B_Q(f_n) r dr \\ &\quad + \int_{y_1}^{y_2} (1 - 2(m + A_\theta[Q]))\Big|\frac{f_n}{r}\Big|^2 r dr.\end{aligned}$$

Since $A_\theta[Q](r) \to 0$ as $r \to 0$, we take $y_1 \to 0$ and $y_2 = r_1$ sufficiently small to obtain

$$\int_{0}^{r_1} \Big|\frac{f_n}{r}\Big|^2 r dr \le |f_n|^2(r_1) + \int_{0}^{r_1} |L_Q f_n|^2 r dr + 2\int_{0}^{r_1} \Big|\frac{f_n}{r}\Big| Q B_Q(|f_n|) r dr.$$

The last term of the RHS can be absorbed into the LHS as r_1 is small. We average in the r_1-variable and take $n \to \infty$ to get

$$\liminf_{n\to\infty} \int_{0}^{r_1} \Big|\frac{f_n}{r}\Big|^2 r dr \lesssim \liminf_{n\to\infty} \int_{r_1}^{2r_1} \Big|\frac{f_n}{r}\Big|^2 r dr. \tag{3.22}$$

From (3.21) and (3.22), we get a contradiction. This proves the claim (3.20).

We are now in position to prove the strong coercivity (3.18). If $m \ge 1$, one can proceed as

$$\begin{aligned}\|L_Q f\|_{L^2} &= \delta\|L_Q f\|_{L^2} + (1-\delta)\|L_Q f\|_{L^2} \\ &\ge \delta(\|\partial_r f\|_{L^2} - C\|r^{-1} f\|_{L^2}) + c(1-\delta)\|r^{-1} f\|_{L^2} \ge \delta\|f\|_{\dot{\mathcal{H}}_m},\end{aligned}$$

provided δ is sufficiently small. If $m = 0$, one uses $|r^{-1} A_\theta[Q]| \lesssim (1+r)^{-1}$ to obtain the desired coercivity. This ends the proof of Lemma 3.9. □

CHAPTER 4

Profile $Q^{(\eta)}$

From now on, we assume $m \geq 1$.

4.1. Role of pseudoconformal phase

In view of pseudoconformal symmetry (1.5), we use the modified profile

$$Q_b(y) := Q(y)e^{-ib\frac{|y|^2}{4}}.$$

We note that for all small $b \geq 0$

$$\|Q_b - Q\|_{L^2} \lesssim \begin{cases} b|\log b|^{\frac{1}{2}} & \text{if } m = 1, \\ b & \text{if } m \geq 2. \end{cases} \tag{4.1}$$

If $m = 1$, the logarithmic factor comes from $\||y|^2 Q\|_{L^2(|y|\lesssim b^{-\frac{1}{2}})} \sim |\log b|^{\frac{1}{2}}$. As we handle the case $m \geq 1$ in the sequel, we only use the upper bound $b|\log b|^{\frac{1}{2}}$. Moreover, as we will be in the regime $b \lesssim \lambda$, we use the upper bound $\lambda|\log\lambda|^{\frac{1}{2}}$. See (6.24), (6.40), and (6.57) for instance.

For a function f and $b \in \mathbb{R}$, recall the notation

$$f_b(y) := f(y)e^{-ib\frac{|y|^2}{4}}.$$

We discuss how the pseudoconformal phase $e^{-ib\frac{|y|^2}{4}}$ acts on the (linearized) evolution of (CSS). We record explicit algebras for later use.

LEMMA 4.1 (Conjugation by pseudoconformal phase). *For any $b \in \mathbb{R}$, we have*

$$L_{f_b}^* \mathbf{D}_+^{(f_b)} f_b = [L_f^* \mathbf{D}_+^{(f)} f]_b + ib\Lambda f_b - ib^2 \partial_b(f_b), \tag{4.2}$$

$$\mathcal{L}_{w_b}\epsilon_b = [\mathcal{L}_w \epsilon]_b + ib\Lambda\epsilon_b - ib^2\partial_b(\epsilon_b), \tag{4.3}$$

for any functions f, w, ϵ.

PROOF. By a direct computation, we get

$$\begin{aligned} \mathbf{D}_+^{(f_b)} g_b &= (\mathbf{D}_+^{(f)} g - ib\tfrac{r}{2} g)_b, \\ L_{f_b} g_b &= (L_f g - ib\tfrac{r}{2} g)_b, \\ L_{f_b}^* g_b &= (L_f^* g + ib\tfrac{r}{2} g)_b, \end{aligned}$$

for any functions f and g. Thus

$$L_{f_b}^* \mathbf{D}_+^{(f_b)} f_b = [(L_f^* + ib\tfrac{r}{2})(\mathbf{D}_+^{(f)} - ib\tfrac{r}{2}) f]_b.$$

Using the identities

$$\begin{aligned} ib\tfrac{r}{2}\mathbf{D}_+^{(f)}f - L_f^*(ib\tfrac{r}{2}f) &= ib\Lambda f, \\ [\Lambda f]_b - \Lambda f_b &= ib\tfrac{r^2}{2}f_b, \\ b^2\tfrac{r^2}{4}f_b &= ib^2\partial_b(f_b), \end{aligned}$$

the first identity (4.2) follows.

To show (4.3), we recall that $\mathcal{L}_w\epsilon$ is the linear part in ϵ of $L_{w+\epsilon}^*\mathbf{D}_+^{(w+\epsilon)}(w+\epsilon)$. Therefore, the identity (4.3) follows by substituting $f = w+\epsilon$ into (4.2) and taking the linear part in ϵ. □

We transfer algebraic identities for Q to Q_b:

LEMMA 4.2 (Algebraic identities from Q_b). *For each $b \in \mathbb{R}$, we have*

$$-L_{Q_b}^*\mathbf{D}_+^{(Q_b)}Q_b + ib\Lambda Q_b = \tfrac{b^2}{4}|y|^2Q_b. \tag{4.4}$$

Moreover, the linearized operator $\mathcal{L}_{Q_b}$ satisfies the following algebraic identities

$$\begin{aligned} \mathcal{L}_{Q_b}iQ_b + b\Lambda Q_b &= b^2\partial_bQ_b, \\ \mathcal{L}_{Q_b}[\Lambda Q]_b - ib\Lambda[\Lambda Q]_b &= -ib^2\partial_b[\Lambda Q]_b, \\ \mathcal{L}_{Q_b}i|y|^2Q_b + b\Lambda(|y|^2Q_b) &= -4i[\Lambda Q]_b + b^2\partial_b(|y|^2Q_b) \\ \mathcal{L}_{Q_b}\rho_b - ib\Lambda\rho_b &= Q_b - ib^2\partial_b\rho_b. \end{aligned}$$

PROOF. Substitute $f = Q$ into (4.2) and use $\mathbf{D}_+^{(Q)}Q = 0$ to get the first identity. To obtain the algebraic identities satisfied by $\mathcal{L}_{Q_b}$, substitute $w = Q$ into (4.3) with help of Proposition 3.4. □

We discuss how the pseudoconformal phase $e^{-ib\frac{|y|^2}{4}}$ represents the pseudoconformal blow-up. Assume that

$$Q_b^\sharp(t,r) = \frac{1}{\lambda(t)}Q_{b(t)}\Big(\frac{r}{\lambda(t)}\Big)e^{i\gamma(t)}$$

solves (CSS). By the dynamical rescaling, Q_b solves

$$i\partial_sQ_b - L_{Q_b}^*\mathbf{D}_+^{(Q_b)}Q_b = i\frac{\lambda_s}{\lambda}\Lambda Q_b + \gamma_sQ_b.$$

Using $i\partial_sQ_b = b_s(\frac{|y|^2}{4}Q_b)$ and (4.4), Q_b satisfies

$$i\Big(\frac{\lambda_s}{\lambda} + b\Big)\Lambda Q_b + \gamma_sQ_b - (b_s+b^2)\tfrac{|y|^2}{4}Q_b = 0. \tag{4.5}$$

Therefore, we obtain the formal parameter equations

$$\frac{\lambda_s}{\lambda} + b = 0, \quad \gamma_s = 0, \quad b_s + b^2 = 0,$$

which are verified by the pseudoconformal blow-up regime

$$\lambda(t) = b(t) = |t| \quad \text{and} \quad \gamma(t) = 0, \qquad \forall t < 0.$$

4.2. A hint to rotational instability

We now discuss the instability mechanism of pseudoconformal blow-up solutions. We first start formal computations that lead to the instability of $S(t)$ or pseudoconformal blow-up solutions. This section assumes the existence of the modified profile $Q^{(\eta)}$ below. In fact, construction of $Q^{(\eta)}$ is an *issue* and will be discussed in the next section.

Fix a small $\eta \geq 0$. We consider a real-valued profile $Q^{(\eta)}(r)$ that modifies $Q = Q^{(0)}$. We let $Q_b^{(\eta)} := e^{-ib\frac{|y|^2}{4}}Q^{(\eta)}$ and

$$Q_b^{(\eta)\sharp}(t,r) := \frac{1}{\lambda(t)}Q_{b(t)}^{(\eta)}\Big(\frac{r}{\lambda(t)}\Big)e^{i\gamma(t)}.$$

Assume that $Q_b^{(\eta)\sharp}$ solves (CSS). Taking the imaginary part of (4.5), we further require

$$\frac{\lambda_s}{\lambda} + b = 0.$$

Now the profile $Q^{(\eta)}$ should satisfy

$$L_{Q^{(\eta)}}^{*}\mathbf{D}_{+}^{(Q^{(\eta)})}Q^{(\eta)} + \gamma_s Q^{(\eta)} - (b_s + b^2)\tfrac{|y|^2}{4}Q^{(\eta)} = 0. \tag{4.6}$$

In case of (NLS), the authors of [**50**] were able to choose[1] $b_s + b^2 = -\eta$ and $\gamma_s = 1$. The instability direction is ρ_{NLS}, which satisfies $\mathcal{L}_{\mathrm{NLS}}\rho_{\mathrm{NLS}} = |y|^2 R$. For (CSS), however,

$$\mathcal{L}_Q(\partial_\eta|_{\eta=0}Q^{(\eta)}) = \tfrac{|y|^2}{4}Q$$

is *not solvable.*[2] In other words, the law $b_s + b^2 = -\eta$ is forbidden. Thus we have to search for a different instability mechanism.

It turns out that there is a *rotational instability* for (CSS). Let us require

$$\gamma_s = \eta \quad \text{and} \quad b_s + b^2 = o(\eta).$$

Assume that there exists a well-decaying solution $Q^{(\eta)}$ to (4.6). Differentiating (4.6) in the η-variable, we have

$$\mathcal{L}_Q(\partial_\eta|_{\eta=0}Q^{(\eta)}) + \eta Q = 0.$$

Thus we may take

$$\partial_\eta|_{\eta=0}Q^{(\eta)} = -\rho,$$

where ρ is constructed in Lemma 3.6.

We now claim that $b_s + b^2$ has nontrivial η^2-order terms as

$$b_s + b^2 = -c\eta^2, \quad c := \frac{1}{(m+1)^2} + o_{\eta\to 0^+}(1).$$

To see this, we take the inner product of (4.6) with $\Lambda Q^{(\eta)}$ to obtain the Pohozaev identity

$$2E[Q^{(\eta)}] + (b_s + b^2)\cdot \tfrac{1}{4}\|yQ^{(\eta)}\|_{L^2}^2 = 0.$$

We can compute the leading order term of $2E[Q^{(\eta)}]$ as

$$2E[Q^{(\eta)}] = \int |\mathbf{D}_{+}^{(Q^{(\eta)})}Q^{(\eta)}|^2 = (\eta^2 + o(\eta^2))\|L_Q\rho\|_{L^2}^2.$$

---

[1] There the sign condition for η is not necessary.

[2] in the sense discussed in Section 3.3.

Therefore,

$$b_s + b^2 = -\frac{4\|L_Q\rho\|_{L^2}^2}{\|rQ\|_{L^2}^2}(\eta^2 + o(\eta^2)).$$

As we know by Remark 3.7 that

$$L_Q\rho = \psi = \frac{1}{2(m+1)} rQ,$$

the claim follows.

Summing up the above discussions, we are led to the ODEs

$$\frac{\lambda_s}{\lambda} + b = 0; \quad \gamma_s = \eta; \quad b_s + b^2 + c\eta^2 = 0. \tag{4.7}$$

There is a global-in-time exact solution

$$\begin{cases} \lambda(t) = \sqrt{t^2 + c\eta^2}, \\ \gamma(t) = \frac{1}{\sqrt{c}} \tan^{-1}(\frac{t}{\eta\sqrt{c}}), \quad \forall t \in \mathbb{R} \\ b(t) = -t. \end{cases}$$

Thus for any fixed $t > 0$, we observe a phase rotation

$$\lim_{\eta \to 0^+} \big(\gamma(t) - \gamma(-t)\big) = \frac{\pi}{\sqrt{c}} = (m+1)\pi.$$

Finally noting that $e^{i\gamma} \cdot f(r)e^{im\theta} = f(r)e^{im(\theta + \frac{\gamma}{m})}$, this corresponds to a spatial rotation with the angle

$$\Big(\frac{m+1}{m}\Big)\pi$$

near time zero.

When $\eta > 0$ is small, $Q_b^{(\eta)\sharp}$ is an exact scattering solution to (CSS), which is close to $S(t)$. At the blow-up time $t = 0$ of $S(t)$, the solution $Q_b^{(\eta)\sharp}$ abruptly takes a spatial rotation by the angle $(\frac{m+1}{m})\pi$. This is a sharp contrast to $S(t)$, which does not rotate at all. We will exhibit the same instability mechanism for other pseudoconformal blow-up solutions.

4.3. Construction of $Q^{(\eta)}$

In this section, we rigorously construct the modified profiles $Q^{(\eta)}$ for all small $\eta > 0$. From (4.6) with (4.7), we expect $Q^{(\eta)}$ to show an exponential decay. If we recall that Q has a polynomial decay, then approximating $Q^{(\eta)}(r)$ by an η-expansion of $Q(r)$ does not make sense for large r. For m small, performing an η-expansion leads to a harmful spatial decay. To resolve this issue, we have to search for a *nonlinear deformation* of $Q^{(\eta)}$. We crucially exploit the *self-duality*. The main novelty of the construction of $Q^{(\eta)}$ is that we can reduce (4.6) to a first-order differential equation. Although we exhibit the same mechanism, we *warn* that $Q^{(\eta)}$ in this section is slightly different from that in Section 4.2. This is merely due to a technical reason.

When $\eta = 0$, we could impose the laws $\gamma_s = 0$ and $b_s + b^2 = 0$, and (4.6) becomes

$$L_Q^* \mathbf{D}_+^{(Q)} Q = 0.$$

Because of the self-duality, this reduces to the Bogomol'nyi equation

$$\mathbf{D}_+^{(Q)} Q = 0.$$

Inspired by this reduction, we twist the Bogomol'nyi equation by the η-parameter that introduces a nontrivial $\gamma_s Q^{(\eta)}$ term in (4.6). This is one of the crucial points that we make use of the self-duality.

LEMMA 4.3 (Modified Bogomol'nyi equation). *Let $\eta \geq 0$. Assume we have a profile $Q^{(\eta)}$ satisfying*

$$\mathbf{D}_+^{(Q^{(\eta)})} P^{(\eta)} = 0, \quad Q^{(\eta)} = e^{-\eta \frac{r^2}{4}} P^{(\eta)},$$

or equivalently,

$$\mathbf{D}_+^{(Q^{(\eta)})} Q^{(\eta)} = -\tfrac{\eta}{2} r Q^{(\eta)}. \tag{4.8}$$

Then $Q^{(\eta)}$ satisfies[3]

$$L_{Q^{(\eta)}}^* \mathbf{D}_+^{(Q^{(\eta)})} Q^{(\eta)} + \eta\theta_\eta Q^{(\eta)} + \eta^2 \tfrac{r^2}{4} Q^{(\eta)} = 0 \tag{4.9}$$

with

$$\theta_\eta := \frac{1}{4\pi} \int |Q^{(\eta)}|^2 - (m+1). \tag{4.10}$$

PROOF. We compute

$$\begin{aligned} L_{Q^{(\eta)}}^* \mathbf{D}_+^{(Q^{(\eta)})} Q^{(\eta)} &= -\frac{\eta}{2} L_{Q^{(\eta)}}^* [rQ^{(\eta)}] \\ &= \frac{\eta}{2}\Big[\Big(\partial_r + \frac{m+1+A_\theta[Q^{(\eta)}]}{r}\Big)(rQ^{(\eta)}) - \Big(\int_r^\infty |Q^{(\eta)}|^2 r' dr'\Big) Q^{(\eta)} \Big] \\ &= \frac{\eta}{2}\Big[r\Big(\mathbf{D}_+^{(Q^{(\eta)})} + \frac{2m+2+2A_\theta[Q^{(\eta)}]}{r}\Big) Q^{(\eta)} - \Big(\int_r^\infty |Q^{(\eta)}|^2 r' dr'\Big) Q^{(\eta)} \Big]. \end{aligned}$$

Using the algebra

$$2A_\theta[Q^{(\eta)}]Q^{(\eta)} - \Big(\int_r^\infty |Q^{(\eta)}|^2 r' dr'\Big) Q^{(\eta)} = -\frac{1}{2\pi}\Big(\int |Q^{(\eta)}|^2\Big) Q^{(\eta)},$$

we have

$$L_{Q^{(\eta)}}^* \mathbf{D}_+^{(Q^{(\eta)})} Q^{(\eta)} = -\eta^2 \frac{r^2 Q^{(\eta)}}{4} + \eta\Big(m+1-\frac{1}{4\pi}\int |Q^{(\eta)}|^2\Big) Q^{(\eta)}.$$

This completes the proof of (4.9). □

In particular,

$$Q_b^{(\eta)\sharp}(t,x) = \frac{1}{\lambda(t)} Q_{b(t)}^{(\eta)}\Big(\frac{x}{\lambda(t)}\Big) e^{i\gamma(t)}$$

is an exact solution to (CSS) if λ, γ, b satisfy

$$\begin{cases} \dfrac{\lambda_s}{\lambda} + b = 0, \\ b_s + b^2 = -\eta^2, \\ \qquad \gamma_s = \eta\theta_\eta. \end{cases} \tag{4.11}$$

[3]Here the profile $Q^{(\eta)}$ is different from that in Section 4.2. Compare (4.11) and (4.7). This is nothing but transforming η to $\eta_1 = \eta_1(\eta)$.

For each fixed $\eta > 0$, the following $(\lambda_\eta, \gamma_\eta, b_\eta)$ solves (4.11):

$$\begin{cases} \lambda_\eta(t) := \langle t \rangle, \\ \gamma_\eta(t) := \theta_\eta \tan^{-1}(\tfrac{t}{\eta}), \\ b_\eta(t) := -t, \end{cases} \qquad \forall t \in \mathbb{R}. \tag{4.12}$$

Since we will construct $Q^{(\eta)}$ as a deformation of Q, we have $\theta_\eta = (m+1) + O(\eta)$.

Now, we solve the first-order equation (4.8) and derive estimates on $Q^{(\eta)}$. We solve (4.8) from $r = 0$ up to $r = R_\eta$ using a perturbative argument around Q. This is possible on the regime where the linearization $Q^{(\eta)} \approx Q + \eta\partial_\eta|_{\eta=0} Q^{(\eta)}$ is valid. The presence of $\eta^2 \frac{|y|^2 Q^{(\eta)}}{4}$ in (4.9) suggests that $Q^{(\eta)}$ has additional exponential decay $e^{-\eta\frac{r^2}{4}}$. Since Q decays polynomially, we expect that the linearization is valid on the region $r \ll \eta^{-\frac{1}{2}}$. Beyond $r = R_\eta$, we cannot view $Q^{(\eta)}$ as a perturbation of Q. Instead, we directly look at (4.8) and observe that $A_\theta[Q^{(\eta)}](R_\eta) \approx -2(m+1)$. This says that $m + A_\theta[Q^{(\eta)}]$ takes negative values. Thus $P^{(\eta)}$ should decay polynomially and can be solved globally on $(0, \infty)$.

For $B > 1$ to be chosen large, let

$$R_\eta := (B\eta)^{-\frac{1}{2}}.$$

PROPOSITION 4.4 (Construction of $Q^{(\eta)}$). *There exist $B > 1$, $\eta^{**} > 0$, and one-parameter families $\{Q^{(\eta)}\}_{\eta \in [0,\eta^{**}]}$ and $\{P^{(\eta)}\}_{\eta \in \{0,\eta^{**}]}$ of smooth real-valued functions on $(0, \infty)$ satisfying the following properties.*

(1) *(Equations of $Q^{(\eta)}$ and $P^{(\eta)}$)*

$$\begin{cases} \mathbf{D}_+^{(Q^{(\eta)})} P^{(\eta)} = 0, \\ Q^{(\eta)} = e^{-\eta\frac{r^2}{4}} P^{(\eta)}, \\ Q^{(0)} = P^{(0)} = Q. \end{cases} \tag{4.13}$$

(2) *(Uniform bounds on $Q^{(\eta)}$) For any $\ell \in \{0, 1\}$, we have*

$$|\partial_r^{(\ell)} Q^{(\eta)}(r)| \lesssim \mathbf{1}_{r \le R_\eta} r^{-\ell} Q + (R_\eta)^{-\ell} Q(R_\eta) \cdot \mathbf{1}_{r \ge R_\eta} e^{-\eta\frac{r^2}{4}}. \tag{4.14}$$

In particular, we have

$$\sup_{\eta \in [0,\eta^{**}]} |\partial_r^{(\ell)} Q^{(\eta)}(r)| \lesssim r^{-\ell} Q. \tag{4.15}$$

(3) *(Differentiability in the η-variable) We have*

$$|Q^{(\eta)} - Q + \eta \cdot (m+1)\rho| \lesssim \eta^2 r^4 Q, \qquad \forall r \le R_\eta. \tag{4.16}$$

(4) *(Subcritical mass of $Q^{(\eta)}$) We have*

$$\int |Q^{(\eta)}|^2 = \int |Q|^2 - \frac{\eta}{2(m+1)} \int |rQ|^2 + o_{\eta \to 0}(\eta). \tag{4.17}$$

In particular,

$$\theta_\eta = (m+1) - \frac{\eta}{8\pi(m+1)} \int |rQ|^2 + o_{\eta \to 0}(\eta). \tag{4.18}$$

(5) *(Decay of $P^{(\eta)}$) We have*

$$\lim_{r \to \infty} \frac{\log P^{(\eta)}}{\log r} = -(m+2) + \frac{\eta}{8\pi(m+1)} \int |rQ|^2 + o_{\eta \to 0}(\eta). \tag{4.19}$$

In particular, the decay of $P^{(\eta)}$ is slower than Q.

(6) *(L^2-difference of $Q^{(\eta)}$ and Q) We have*

$$\|Q^{(\eta)} - Q\|_{L^2} \lesssim \begin{cases} \eta|\log \eta|^{\frac{1}{2}} & \text{if } m = 1, \\ \eta & \text{if } m \geq 2. \end{cases} \tag{4.20}$$

PROOF. The small parameter $\eta^{**} > 0$ will be chosen later. Let us assume $\eta^{**} \leq 1$ for now. We will introduce a renormalized unknown $\widetilde{v}$ and write the equation in $\widetilde{v}$. To get some motivation, we first note a formal computation. If $Q^{(\eta)}$ exists, then we have

$$\mathbf{D}_+^{(Q^{(\eta)})} Q^{(\eta)} = -\eta \tfrac{r}{2} Q^{(\eta)}.$$

Differentiating in the η-variable at 0, we get the relation

$$L_Q[\partial_\eta Q^{(\eta)}|_{\eta=0}] = -\tfrac{1}{2} rQ.$$

Recalling $L_Q \rho = \psi = \frac{1}{2(m+1)} rQ$ given in Remark 3.7, we may set

$$\partial_\eta Q^{(\eta)}|_{\eta=0} = -(m+1)\rho.$$

We now introduce the unknown

$$v := \eta^{-2}(Q^{(\eta)} - Q + \eta \cdot (m+1)\rho).$$

In terms of v, the equation (4.13) reads

$$\begin{aligned} L_Q v = {} & \big(\tfrac{m+1}{2} r\rho - (m+1)^2 \rho B_Q \rho\big) \\ & + \eta\big(-\tfrac{1}{2} r + (m+1) B_Q \rho\big) v + \eta^2 \big(\tfrac{1}{2} B_Q v\big) v. \end{aligned}$$

As in the proof of Lemma 3.6, we introduce the renormalized unknown

$$\widetilde{v} := Q^{-1} v$$

and write the integral equation

$$\begin{aligned} \widetilde{v}(r) = {} & \int_0^r \big(\tfrac{m+1}{2} \cdot \frac{r'\rho}{Q} - (m+1)^2 \frac{\rho}{Q} B_Q \rho\big) dr' + \int_0^r \Big(\int_0^{r'} Q^2 \widetilde{v} r'' dr''\Big) \frac{dr'}{r'} \\ & + \eta \int_0^r \big(-\tfrac{1}{2} r' + (m+1) B_Q \rho\big) \widetilde{v} dr' + \eta^2 \int_0^r \big(\tfrac{1}{2} B_Q v\big) \widetilde{v} dr'. \end{aligned} \tag{4.21}$$

It is now natural to expect that $\widetilde{v}(r) \lesssim r^4$. By a simple contraction argument, we can choose small $r_0 > 0$ and large $C_0 > 0$ independent of η such that $\widetilde{v}(r) \leq C_0 r^4$ for all $r \in [0, r_0]$ and $\eta \in [0, 1]$.

We now construct $\widetilde{v}$ beyond $[0, r_0)$. For $\delta > 0$ to be chosen small later independent of η, choose an increasing sequence $\{r_n\}_{n \geq 0} \subset \mathbb{R}$ such that $r_n \to \infty$, $r_{n+1} \leq 2r_n$, and $\int_{r_n}^{r_{n+1}} Q^2 r'' dr'' \leq \delta$. Set $\eta_0 = 1$. We follow the proof of Lemma 3.6 with an extra care on the smallness of η. Indeed, if $\delta > 0$ is chosen sufficiently small, then we can inductively construct sequences of $\{\eta_n\}_{n \geq 0}$, $\{C_n\}_{n \geq 0}$, and $\widetilde{v}(r)$ on $[0, r_n]$: assume we have constructed $\widetilde{v}(r)$ on $[0, r_n]$ for all $\eta \in [0, \eta_n]$ such that $|\widetilde{v}(r)| \leq C_n r^4$ on $[0, r_n]$ for all η. Then, there exist $0 < \eta_{n+1} \leq \eta_n$ and $C_{n+1} \geq C_n$ such that we can construct $\widetilde{v}(r)$ on $[r_n, r_{n+1}]$ for all $\eta \in [0, \eta_{n+1}]$ such that $|\widetilde{v}(r)| \leq C_{n+1} r^4$ on $[0, r_{n+1}]$ for all η. It is important that C_n is independent of η, once η is smaller than η_n.

Now we claim that after finite steps we can further construct $\widetilde{v}$ on $[0, R_\eta]$. In what follows, we only show a bound on $\widetilde{v}$ via a bootstrap argument. The existence of $\widetilde{v}$ automatically follows by a standard arguemnt. We claim that there exist large

$N \in \mathbb{N}$, $B > 1$, $C > C_N$, and small $\eta^{**} \in (0, \eta_N]$ such that we can construct $\widetilde{v}$ with the bound $|\widetilde{v}(r)| \leq Cr^4$ on $[0, R_\eta]$ for all $\eta \in [0, \eta^{**}]$. To show this claim, it suffices to show the bound $|\widetilde{v}(r)| \leq Cr^4$ for $r \in [r_N, R_\eta]$. We simply write the equation (4.21) as

$$\widetilde{v}(r) = O(r^4) + \int_0^r \Big(\int_0^{r'} Q^2 \widetilde{v} r'' dr'' \Big) \frac{dr'}{r'} + \eta \int_0^r O(r') \widetilde{v} dr' + \eta^2 \int_0^r (\tfrac{1}{2} B_Q v) \widetilde{v} dr'.$$

We consider the bootstrap hypothesis $|\widetilde{v}(r)| \leq Cr^4$. Using the estimate $\int_1^r Q^2 (r')^5 dr' \lesssim \log r$ for $r \geq r_N$, we get

$$\widetilde{v}(r) \lesssim r^4 + C \log^2 r + C\eta r^6 + C^2 \eta^2 r^4 \log r$$

for $r \geq r_N$. If we choose N large such that $\log^2 r \ll r^4$, $C > C_N$ large, and $\eta^{**} \in (0, \eta_N]$ small such that $C\eta^{**} \ll 1$, then we have

$$\widetilde{v}(r) \leq (\tfrac{1}{2} + \eta \log r + \eta r^2) \cdot Cr^4$$

for $r \geq r_N$. Therefore, if we choose $B > 1$ sufficiently large, then we have

$$\widetilde{v}(r) \leq \tfrac{3}{4} Cr^4$$

for all $r \in [r_N, R_\eta]$. By a continuity argument, the claim follows. Tracking back the definitions of $\widetilde{v}$ and v, the proof of (4.16) is completed.

We have constructed $Q^{(\eta)}$ on $(0, R_\eta]$ for all $\eta \in [0, \eta^{**}]$. By (4.16), we have

$$\begin{aligned} \big(A_\theta[Q^{(\eta)}] - A_\theta[Q]\big)(R_\eta) &= \frac{1}{4\pi} \int \mathbf{1}_{r \leq R_\eta} (|Q|^2 - |Q^{(\eta)}|^2) \\ &= \eta \frac{(m+1)}{2\pi} \Big(\int \mathbf{1}_{r \leq R_\eta} Q\rho \Big) + o_{\eta \to 0}(\eta) \\ &= \eta \frac{(m+1)}{2\pi} (Q, \rho)_r + o_{\eta \to 0}(\eta). \end{aligned}$$

We then apply $(Q, \rho)_r = (L_Q^* L_Q \rho, \rho)_r = \|L_Q \rho\|_{L^2}^2$, Remark 3.6, and $A_\theta[Q](+\infty) = -2(m+1)$ to obtain

$$A_\theta[Q^{(\eta)}](R_\eta) = -2(m+1) + \frac{\eta}{8\pi(m+1)} \|rQ\|_{L^2}^2 + o_{\eta \to 0}(\eta). \tag{4.22}$$

We now construct $Q^{(\eta)}$ beyond $r = R_\eta$. On the region $r \gtrsim R_\eta$, the linear approximation of $Q^{(\eta)}$ is not accurate. We consider the equation of $P^{(\eta)}$ instead:

$$\partial_r P^{(\eta)} = \frac{m + A_\theta[Q^{(\eta)}]}{r} P^{(\eta)}.$$

By (4.22), we already have $A_\theta[Q^{(\eta)}](R_\eta) \leq -2m - \frac{3}{2}$ for all $\eta \in [0, \eta^{**}]$. By the comparison principle, we get

$$P^{(\eta)}(r) \leq P^{(\eta)}(R_\eta) \Big(\frac{r}{R_\eta} \Big)^{-m-\frac{3}{2}} \tag{4.23}$$

for all $r \geq R_\eta$. This shows that $P^{(\eta)}$ (and $Q^{(\eta)}$) exists globally on $(0, \infty)$.

We now show (4.14) and (4.15). Since (4.15) easily follows from (4.14), we focus on (4.14). Note that (4.16) says that $Q^{(\eta)}(r) \sim Q(r)$ for $r \leq R_\eta$. This verifies (4.14) with $\ell = 0$ on the region $r \leq R_\eta$. On the region $r \geq R_\eta$, we use the exponential decay $e^{-\eta \frac{r^2}{4}}$ of $Q^{(\eta)}$ and (4.23) to conclude (4.14) with $\ell = 0$ on the region $r \geq R_\eta$.

Thus (4.14) hold when $\ell = 0$. We note that (4.14) with $\ell = 1$ follows by applying the equation

$$\partial_r Q^{(\eta)} = -\frac{m + A_\theta[Q^{(\eta)}]}{r} Q^{(\eta)} - \frac{1}{2}\eta r^2 Q^{(\eta)}.$$

We turn to show (4.17), (4.18), and (4.19). Note that (4.15) yields

$$\|\mathbf{1}_{r>R_\eta} Q^{(\eta)}\|_{L^2} = o_{\eta\to 0}(\eta).$$

Substituting this into (4.22) gives (4.17). To prove (4.18), substitute (4.17) into (4.10). To prove (4.19), observe that

$$\partial_r \log P^{(\eta)} = \frac{m + A_\theta[Q^{(\eta)}]}{r}.$$

Therefore, the decay rate of $P^{(\eta)}$ is obtained from the quantity $m + A_\theta[Q^{(\eta)}](+\infty)$, which is $-(m+2) + \frac{\eta}{8\pi(m+1)}\|rQ\|_{L^2}^2$. This completes the proof of (4.19).

Finally, we show (4.20). Note that

$$\begin{aligned}\|Q^{(\eta)} - Q\|_{L^2} &= \|\mathbf{1}_{r\le R_\eta}(Q^{(\eta)} - Q)\|_{L^2} + o_{\eta\to 0}(\eta) \\ &= \eta(m+1)\|\mathbf{1}_{r\le R_\eta}\rho\|_{L^2} + o_{\eta\to 0}(\eta).\end{aligned}$$

From the asymptotics (3.16) of ρ, (4.20) follows. □

We record identities induced by the phase and scaling symmetries on (4.9).

LEMMA 4.5 (Algebraic Identities). *We have*

$$\mathcal{L}_{Q^{(\eta)}} iQ^{(\eta)} + \eta\theta_\eta iQ^{(\eta)} + \eta^2 \tfrac{|y|^2}{4} iQ^{(\eta)} = 0, \tag{4.24}$$

$$\mathcal{L}_{Q^{(\eta)}} \Lambda Q^{(\eta)} + \eta\theta_\eta \Lambda Q^{(\eta)} + \eta^2 \tfrac{|y|^2}{4} \Lambda Q^{(\eta)} = -2\eta\theta^{(\eta)} Q^{(\eta)} - 4\eta^2 \tfrac{|y|^2}{4} Q^{(\eta)}. \tag{4.25}$$

PROOF. One can obtain these identities by differentiating the phase and scaling symmetries at $Q^{(\eta)}$. Differentiating

$$L^*_{e^{ia}Q^{(\eta)}} \mathbf{D}_+^{(e^{ia}Q^{(\eta)})} e^{ia} Q^{(\eta)} + \eta\theta_\eta e^{ia} Q^{(\eta)} + \eta^2 \tfrac{|y|^2}{4} e^{ia} Q^{(\eta)} = 0$$

in the a-variable at $a = 0$, we obtain (4.24). To obtain (4.25), we temporarily write $f_\lambda(y) := \frac{1}{\lambda} f(\frac{y}{\lambda})$ and consider

$$\begin{aligned} 0 &= [L^*_{Q^{(\eta)}} \mathbf{D}_+^{(Q^{(\eta)})} Q^{(\eta)} + \eta\theta_\eta Q^{(\eta)} + \eta^2 \tfrac{|y|^2}{4} Q^{(\eta)}]_\lambda \\ &= \lambda^2 L^*_{Q_\lambda^{(\eta)}} \mathbf{D}_+^{(Q_\lambda^{(\eta)})} Q_\lambda^{(\eta)} + \eta\theta_\eta Q_\lambda^{(\eta)} + \eta^2 \lambda^{-2} \tfrac{|y|^2}{4} Q_\lambda^{(\eta)} \end{aligned}$$

for $\lambda \in (0,\infty)$. Differentiating this in the λ-variable at $\lambda = 1$, we get

$$\begin{aligned} &\mathcal{L}_{Q^{(\eta)}} \Lambda Q^{(\eta)} + \eta\theta_\eta \Lambda Q^{(\eta)} + \eta^2 \tfrac{|y|^2}{4} \Lambda Q^{(\eta)} \\ &\quad = 2L^*_{Q^{(\eta)}} \mathbf{D}_+^{(Q^{(\eta)})} Q^{(\eta)} - 2\eta^2 \tfrac{|y|^2}{4} Q^{(\eta)} = -2\eta\theta_\eta Q^{(\eta)} - 4\eta^2 \tfrac{|y|^2}{4} Q^{(\eta)}. \end{aligned}$$

This completes the proof of (4.25). □

CHAPTER 5

Setup for modulation analysis

In this chapter, we will reduce Theorems 1.1 and 1.3 to the main bootstrap Lemma 5.3. We will introduce time-dependent modulation parameters (λ, γ, b). Here λ, γ, and b correspond to the scaling, phase, and pseudoconformal phase parameters, respectively. We write the blow-up ansatz in terms of (λ, γ, b) and the decomposition into the modified profile $Q^{(\eta)}$, some small asymptotic profile z^*, and error ϵ. We first determine the evolution equation of z. We then fix dynamical laws of (λ, γ, b) to grant the desired pseudoconformal blow-up rate. The main novelty of this step is to detect the strong interactions between $Q_b^\sharp$ and z, and incorporate them into our blow-up ansatz.

5.1. Corrections from the interaction between $Q_b^\sharp$ and z

Our goal is to construct pseudoconformal blow-up solution with given m-equivariant asymptotic profile z^*. We will require that z^* is small in some Sobolev space and is degenerate at the origin. Given z^*, we fix the backward-in-time evolution of $z(t)$ with $z(0) = z^*$. Near the blow-up time, $Q_b^\sharp$ is concentrated at the origin but z is a regular solution; they are decoupled in scales. Thus when one presumes that $Q_b^\sharp + z$ solves (CSS), one might expect that the mixed term of the nonlinearity becomes negligible. This is indeed true for local nonlinearities, for example $|u|^2 u$. In the context of (NLS), Bourgain-Wang [**3**] and Merle-Raphaël-Szeftel [**50**] make use of this observation, and hence there is no strong influence between $Q_b^\sharp$ and z. Thus it is enough to evolve $z(t)$ by (NLS) itself. In (CSS), it turns out that there are strong interactions between $Q_b^\sharp$ and z due to the long-range interactions in A_θ and A_0. We have to capture strong interactions and incorporate the effects into the decomposition of our blow-up ansatz. More precisely, we will modify the z-evolution and add a correction to the γ-evolution.

We first explain the heuristics of our argument. Assume that $Q_b^\sharp + z$ solves (CSS) (equivalently, (1.28)) in the sense that

$$0 \approx (5.1),$$

where

$$\begin{aligned} & i\partial_t (Q_b^\sharp + z) - L^*_{Q_b^\sharp + z} \mathbf{D}_+^{(Q_b^\sharp + z)} (Q_b^\sharp + z) \\ & = \begin{cases} (i\partial_t + \partial_{rr} + \frac{1}{r}\partial_r)(Q_b^\sharp + z) - \Big(\frac{m + A_\theta[Q_b^\sharp + z]}{r}\Big)^2 (Q_b^\sharp + z) \\ - A_0[Q_b^\sharp + z](Q_b^\sharp + z) + |Q_b^\sharp + z|^2 (Q_b^\sharp + z) \end{cases} \end{aligned} \tag{5.1}$$

We assume the decoupling in scales, for example

$$A_\theta[Q_b^\sharp + z] \approx A_\theta[Q_b^\sharp] + A_\theta[z] \quad \text{and} \quad |Q_b^\sharp + z|^2 \approx |Q_b^\sharp|^2 + |z|^2,$$

but $A_\theta[Q_b^\sharp]z$ is not negligible since $A_\theta[Q_b^\sharp](r) \to -2(m+1)$ as $r \to \infty$. We then obtain

$$0 \approx (5.2),$$

where

(5.2)
$$\begin{cases} (i\partial_t + \partial_{rr} + \frac{1}{r}\partial_r)(Q_b^\sharp + z) - \Big(\frac{m + A_\theta[Q_b^\sharp] + A_\theta[z]}{r}\Big)^2 (Q_b^\sharp + z) \\ \quad + \Big(\int_r^\infty (m + A_\theta[Q_b^\sharp] + A_\theta[z])(|Q_b^\sharp|^2 + |z|^2)\frac{dr'}{r'}\Big)(Q_b^\sharp + z) + |Q_b^\sharp|^2 Q_b^\sharp + |z|^2 z. \end{cases}$$

Separately collecting the evolutions for $Q_b^\sharp$ and z, and further applying the decoupling in scales, we have

$$0 \approx (5.3) + (5.4),$$

where

$$\begin{cases} (i\partial_t + \partial_{rr} + \frac{1}{r}\partial_r)Q_b^\sharp - \Big(\frac{m + A_\theta[Q_b^\sharp]}{r}\Big)^2 Q_b^\sharp \\ \quad - A_0[Q_b^\sharp]Q_b^\sharp + \Big(\int_r^\infty (m + A_\theta[Q_b^\sharp] + A_\theta[z])|z|^2\frac{dr'}{r'}\Big)Q_b^\sharp + |Q_b^\sharp|^2 Q_b^\sharp \end{cases} \tag{5.3}$$

$$\begin{cases} (i\partial_t + \partial_{rr} + \frac{1}{r}\partial_r)z - \Big(\frac{m + A_\theta[Q_b^\sharp] + A_\theta[z]}{r}\Big)^2 z \\ \quad + \Big(\int_r^\infty (m + A_\theta[Q_b^\sharp] + A_\theta[z])|z|^2\frac{dr'}{r'}\Big)z + |z|^2 z. \end{cases} \tag{5.4}$$

The above suggests us to evolve z by setting (5.4) equal to zero. Observing that $Q_b^\sharp$ is concentrated at the origin, z feels $Q_b^\sharp$ as a point charge. Thus we further approximate $A_\theta[Q_b^\sharp](r)$ by its value at spatial infinity $A_\theta[Q_b^\sharp](\infty) = -2(m+1)$. Then one can view (5.4) as (CSS) under $-(m+2)$-equivariance. For (5.3), in view of decoupling in scales, we again replace $A_\theta[Q_b^\sharp]$ by its value at spatial infinity and $\int_r^\infty$ by $\int_0^\infty$. As a result, we have

$$0 \approx (5.5) + (5.6),$$

where

$$\begin{cases} (i\partial_t + \partial_{rr} + \frac{1}{r}\partial_r)Q_b^\sharp - \Big(\frac{m + A_\theta[Q_b^\sharp]}{r}\Big)^2 Q_b^\sharp \\ \quad - A_0[Q_b^\sharp]Q_b^\sharp - \theta_{z\to Q_b^\sharp} Q_b^\sharp + |Q_b^\sharp|^2 Q_b^\sharp, \end{cases} \tag{5.5}$$

$$\begin{cases} (i\partial_t + \partial_{rr} + \frac{1}{r}\partial_r)z - \Big(\frac{-m - 2 + A_\theta[z]}{r}\Big)^2 z \\ \quad + \Big(\int_r^\infty (-m - 2 + A_\theta[z])|z|^2\frac{dr'}{r'}\Big)z + |z|^2 z, \end{cases} \tag{5.6}$$

and

$$\theta_{z\to Q_b^\sharp} := -\int_0^\infty (-m - 2 + A_\theta[z])|z|^2 \frac{dr'}{r'}. \tag{5.7}$$

So far, we have discussed assuming $\eta = 0$. In case of $\eta > 0$, we perform the same argument with a suitable modification, replacing $Q_b^\sharp$ by $Q_b^{(\eta)\sharp}$. However, in the z-evolution, one cannot view (5.4) as (CSS) under $-(m+2)$-equivariance because of $A_\theta[Q^{(\eta)\sharp}](\infty) \neq -2(m+1)$. Nevertheless, as we know $A_\theta[Q^{(\eta)\sharp}](\infty) =$

$-2(m+1)+O(\eta)$ by (4.17), we evolve z the same as in the case $\eta=0$. Later, we have to handle the error from $A_\theta[Q^{(\eta)\sharp}]-A_\theta[Q^\sharp]$. On the other hand, the analogue of $\theta_{z\to Q_b^\sharp}$ in the case of $\eta>0$ is

$$\theta_{z\to Q_b^{(\eta)\sharp}} := -\int_0^\infty (m+A_\theta[Q^{(\eta)\sharp}](\infty)+A_\theta[z])|z|^2\frac{dr'}{r'}. \tag{5.8}$$

REMARK 5.1. In the above z-evolution, it is important that $|m+A_\theta[Q](\infty)|-m=2$ is a nonnegative even integer. Indeed, assume that z evolves under

$$i\partial_t z+\Big(\partial_{rr}+\frac{1}{r}\partial_r-\frac{k^2}{r^2}\Big)z\approx 0$$

for some $k\in\mathbb{R}$. Then a generic solution z would have $z(t,r)\sim r^{|k|}$ at the origin. In view of Lemma A.3, if $|k|-m$ is not a nonnegative even integer, then it is absurd to see that $z(t,x)=z(t,r)e^{im\theta}$ is a smooth m-equivariant solution. In other words, if we evolve z using $m+A_\theta[Q^{(\eta)\sharp}](\infty)$, we cannot have sufficient smoothness of z, especially at the origin. Note that from (1.21) and (1.27), $m+A_\theta[Q](\infty)$ corresponds to the decay rate of $Q\sim r^{-(m+2)}\sim r^{(m+A_\theta[Q](\infty))}$ as $r\to\infty$. In other words, it is important that the decay rate of the m-equivariant static solution Q is $m+2\ell$ for some nonnegative integer ℓ.

Motivated from (5.6), we define $\widetilde{z}(t,x):=z(t,x)e^{-i(2m+2)\theta}$ and evolve $\widetilde{z}$ under the $-(m+2)$-*equivariant* (CSS)

$$i\partial_t\widetilde{z}-L_{\widetilde{z}}^*\mathbf{D}_+^{(\widetilde{z})}\widetilde{z}=0. \tag{5.9}$$

Then, (5.9) is equivalent to that the m-equivariant solution z evolves under the *(CSS) with the external potential*

$$\begin{cases} i\partial_t z-L_z^*\mathbf{D}_+^{(z)}z=V_{Q_b^\sharp\to z}z,\\ z(0,x)=z^*(x),\end{cases} \tag{zCSS}$$

where z^* is a small profile and $V_{Q_b^\sharp\to z}$ is the external potential defined by

$$V_{Q_b^\sharp\to z}:=\frac{4(m+1)}{r^2}-\frac{4(m+1)}{r^2}A_\theta[z]+\Big(\int_r^\infty 2(m+1)|z|^2\frac{dr'}{r'}\Big). \tag{5.10}$$

By Proposition 2.9, have the standard L^2-critical local well-posedness, and small data global well-posedness and scattering for (5.9) under the $-(m+2)$-equivariant symmetry.

We prepare small $-(m+2)$-equivariant data $\widetilde{z}^*$ and denote by $\widetilde{z}$ the corresponding global $-(m+2)$-equivariant solution $\widetilde{z}$. More precisely, we assume $\|\widetilde{z}\|_{H^k_{-(m+2)}}<\alpha^*$ for some $k=k(m)$. In view of Lemma A.8, we may choose $k=k(m)>m+3$ to have a degeneracy bound of $\widetilde{z}$ near the origin:

$$\sup_{t\in[-1,0]}\big(|\partial_r^{(\ell)}\widetilde{z}(t,r)|+\tfrac{1}{r^\ell}|\widetilde{z}(t,r)|\big)\lesssim\alpha^*\begin{cases} r^{m+2-\ell} & \text{if } r\le 1,\\ 1 & \text{if } r\ge 1,\end{cases}\qquad \forall\ell\in\{0,1\}.$$

This bound easily transfers to $z:=e^{i(2m+2)\theta}\widetilde{z}$. Note that if $\widetilde{z}$ is smooth $-(m+2)$-equivariant, then z is smooth m-equivariant.

Therefore, we let z^* satisfy the assumption (H) *with sufficiently small α^*, and $z:\mathbb{R}\times\mathbb{R}^2\to\mathbb{C}$ be an m-equivariant solution to* (zCSS) *with the initial data z^*. Then z satisfies the following properties:*

(1) z is global and scatters under the evolution $i\partial_t + \Delta_{-m-2}$.[1]
(2) (Strichartz bound) $\|z\|_{L_t^p L_x^q} \lesssim \alpha^*$ for any admissible pairs (p,q).
(3) (Energy bound) $\|z\|_{L_t^\infty H_x^k} \lesssim \alpha^*$.
(4) (Vanishing at origin)

$$\sup_{t\in[-1,0]} \big(|\partial_r^{(\ell)} z(t,r)| + \tfrac{1}{r^\ell}|z(t,r)|\big) \lesssim \alpha^* \cdot \begin{cases} r^{m+2-\ell} & \text{if } r \le 1, \\ 1 & \text{if } r \ge 1, \end{cases} \qquad \forall \ell \in \{0,1\}.$$

5.2. Evolution of ϵ

We begin the modulation analysis. For a small fixed $\eta \ge 0$, we write a solution $u^{(\eta)}$ of the form

$$u^{(\eta)}(t,x) = \frac{e^{i\gamma(t)}}{\lambda(t)}(Q_{b(t)}^{(\eta)} + \epsilon)(t, \frac{x}{\lambda(t)}) + z(t,x). \tag{5.11}$$

Here z is a small global solution to (zCSS) with $z(0) = z^*$ as in Section 5.1. We want to construct a family of solutions $u^{(\eta)}$ such that $u^{(0)}$ blows up at $t = 0$ with the pseudoconformal rate and for $\eta > 0$, $u^{(\eta)}$ is a global scattering solution with $u^{(\eta)} \to u^{(0)}$ as $\eta \to 0$. More precisely, we expect our modulation parameters to satisfy

$$(b(t), \lambda(t), \gamma(t)) \approx \begin{cases} (-t, |t|, -\frac{m+1}{2}\pi) & \text{if } \eta = 0, \\ (-t, \langle t\rangle, \theta_\eta \tan^{-1}(\frac{t}{\eta})) & \text{if } \eta > 0, \end{cases}$$

for t near zero ($t < 0$ when $\eta = 0$).

If one is only interested in the construction of a blow-up solution, one may disregard η and only work with the case $\eta = 0$. See Remark 5.5. In our presentation, we first derive an equation of ϵ in the case $\eta = 0$ for readability. And then, we will point out what are changed when $\eta > 0$.

Note that we haven't specified the choice of the modulation parameters $b(t)$, $\lambda(t), \gamma(t)$ at each time t. We will fix them by imposing two orthogonality conditions and one relation between b and λ in Section 5.3. In this section, we focus on deriving the evolution equation for ϵ without fixing the laws of b, λ, and γ.

Recall Section 2.5 on $\sharp$ and $\flat$ operations. In (5.11), the functions u and z are defined with the (t,x)-variables, but Q_b and ϵ are functions of (s,y). Thus by $\flat$ operation one can change u and z to functions of (s,y), i.e. $u^\flat(s,y)$ and $z^\flat(s,y)$. Similarly by $\sharp$ operation, one can change Q_b and ϵ to functions of (t,x), i.e. $Q_{b(t)}^\sharp(x)$ and $\epsilon^\sharp(t,x)$. In the following, we take $\flat$ or $\sharp$ operations to convert equations in the (t,x)-variables and in the (s,y)-variables.

In view that Q_b and z live on different scales, it is convenient to introduce general notations to describe their interactions. Consider two functions f and g such that f has shorter length scale than g. f and g can be functions of either (s,y) or (t,x). For example, we may consider $(f,g) = (Q_b^\sharp, z)$ or $(Q_b, z^\flat)$. Firstly, we use

$$R_{f,g} := L_{f+g}^* \mathbf{D}_+^{(f+g)}(f+g) - L_f^* \mathbf{D}_+^{(f)} f - L_g^* \mathbf{D}_+^{(g)} g,$$

[1] z also scatters under the usual evolution $i\partial_t + \Delta_m$. See Remark 5.6.

which is the interaction term between f and g. We will use this for $R_{Q_b,z^\flat}$, $R_{Q_b^\sharp,z}$, and so on. We will use notations

$$\begin{aligned} V_{f\to g} &:= \frac{1}{r^2}\Big((m+A_\theta[f](+\infty)+A_\theta[g])^2-(m+A_\theta[g])^2\Big) \\ &\quad -\int_r^\infty A_\theta[f](+\infty)|g|^2\frac{dr'}{r'}, \\ \theta_{g\to f} &:= -\int_0^\infty (m+A_\theta[f](\infty)+A_\theta[g])|g|^2\frac{dr'}{r'}. \end{aligned}$$

The strong interaction from f to g is given by $V_{f\to g}g$, and that from g to f is given by $\theta_{g\to f}f$. We denote the marginal interaction by

$$\widetilde{R}_{f,g} := R_{f,g} - V_{f\to g}g - \theta_{g\to f}f.$$

Note that

$$[R_{f,g}]^\sharp = \lambda^2[R_{f^\sharp,g^\sharp}], \quad [V_{f\to g}g]^\sharp = \lambda^2 V_{f^\sharp\to g^\sharp}g^\sharp, \quad \theta_{g\to f} = \lambda^2\theta_{g^\sharp\to f^\sharp}.$$

In case of $(f,g) = (Q_b^{(\eta)}, z^\flat)$, we will denote

$$\widetilde{\gamma}_s := \gamma_s + \theta_{z^\flat\to Q_b^{(\eta)}} = \gamma_s + \lambda^2\theta_{z\to Q_b^{(\eta)\sharp}}. \tag{5.12}$$

Next, for functions w and ϵ, we denote by

$$R_{(w+\epsilon)-w} := L_{w+\epsilon}^*\mathbf{D}_+^{(w+\epsilon)}(w+\epsilon) - L_w^*\mathbf{D}_+^{(w)}w - \mathcal{L}_w\epsilon$$

the quadratic and higher order terms in ϵ.

The case $\eta = 0$. Here, we assume $\eta = 0$ and derive the equation of ϵ. If we take the $\flat$ operation to (CSS), then by Section 2.5,

$$\begin{aligned} i\partial_t u - L_u^*\mathbf{D}_+^{(u)}u &= 0, \\ i\partial_s u^\flat - L_{u^\flat}^*\mathbf{D}_+^{(u^\flat)}u^\flat &= i\frac{\lambda_s}{\lambda}\Lambda u^\flat + \gamma_s u^\flat. \end{aligned}$$

The evolution of Q_b is given by Lemma 4.2

$$\begin{aligned} & i\partial_s Q_b - L_{Q_b}^*\mathbf{D}_+^{(Q_b)}Q_b + ib\Lambda Q_b = (b_s+b^2)\tfrac{|y|^2}{4}Q_b, \\ & \quad i\partial_t Q_b^\sharp - L_{Q_b^\sharp}^*\mathbf{D}_+^{(Q_b^\sharp)}Q_b^\sharp = -i\frac{1}{\lambda^2}\Big(\frac{\lambda_s}{\lambda}+b\Big)\Lambda Q_b^\sharp - \frac{\gamma_s}{\lambda^2}Q_b^\sharp + \frac{1}{\lambda^2}(b_s+b^2)[\tfrac{|y|^2}{4}Q]_b^\sharp. \end{aligned}$$

Here, we remark that $\Lambda Q_b = [\Lambda Q]_b - ib\frac{|y|^2}{2}Q_b$ and $\Lambda Q_b^\sharp = [\Lambda Q]_b^\sharp - ib[\frac{|y|^2}{2}Q]_b^\sharp$.

Define

$$w(t,x) := Q_b^\sharp(t,x) + z(t,x) = u(t,x) - \epsilon^\sharp(t,x).$$

Viewing $w = Q_b^\sharp + z$, by (zCSS), w solves

$$\begin{aligned} i\partial_t w - L_w^*\mathbf{D}_+^{(w)}w &= -i\frac{1}{\lambda^2}\Big(\frac{\lambda_s}{\lambda}+b\Big)\Lambda Q_b^\sharp - \frac{\gamma_s}{\lambda^2}Q_b^\sharp + \frac{1}{\lambda^2}(b_s+b^2)[\tfrac{|y|^2}{4}Q]_b^\sharp \\ &\quad + (V_{Q_b^\sharp\to z}z - R_{Q_b^\sharp,z}). \end{aligned}$$

Taking the $\flat$ operation, $w^\flat$ solves

$$\begin{aligned} i\partial_s w^\flat - L_{w^\flat}^*\mathbf{D}_+^{(w^\flat)}w^\flat + ib\Lambda w^\flat &= i\Big(\frac{\lambda_s}{\lambda}+b\Big)\Lambda z^\flat + \gamma_s z^\flat + (b_s+b^2)\tfrac{|y|^2}{4}Q_b \\ &\quad + (V_{Q_b\to z^\flat}z^\flat - R_{Q_b,z^\flat}). \end{aligned}$$

Recall that the external potential term $V_{Q_b^\sharp \to z}$ is introduced to capture strong interactions in $R_{Q_b^\sharp, z}$. We also note that $V_{Q_b^\sharp \to z}$ does not depend on modulation parameters, but only on z.

Finally we derive the equation of ϵ. Subtracting the equation of $u^\flat$ by that of $w^\flat$, we have

$$\begin{aligned} i\partial_s \epsilon - \mathcal{L}_{w^\flat}\epsilon + ib\Lambda\epsilon = i\Big(\frac{\lambda_s}{\lambda} + b\Big)\Lambda(Q_b + \epsilon) + \widetilde{\gamma}_s Q_b + \gamma_s \epsilon - (b_s + b^2)\tfrac{|y|^2}{4} Q_b \\ + \widetilde{R}_{Q_b, z^\flat} + R_{u^\flat - w^\flat}. \end{aligned}$$

By taking the $\sharp$ operation, we have

$$\begin{aligned} i\partial_t \epsilon^\sharp - \mathcal{L}_w \epsilon^\sharp = \frac{1}{\lambda^2} i\Big(\frac{\lambda_s}{\lambda} + b\Big)\Lambda Q_b^\sharp + \frac{1}{\lambda^2}\widetilde{\gamma}_s Q_b^\sharp - \frac{1}{\lambda^2}(b_s + b^2)[\tfrac{|y|^2}{4} Q]_b^\sharp \\ + \widetilde{R}_{Q_b^\sharp, z} + R_{u-w}. \end{aligned}$$

The case $\eta > 0$. We now consider the case $\eta > 0$. Recall (4.9) and (4.2) as

$$i\partial_s Q_b^{(\eta)} - L_{Q_b^{(\eta)}}^* \mathbf{D}_+^{(Q_b^{(\eta)})} Q_b^{(\eta)} + ib\Lambda Q_b^{(\eta)} - \eta\theta_\eta Q_b^{(\eta)} = (b_s + b^2 + \eta^2)\tfrac{|y|^2}{4} Q_b^{(\eta)}.$$

Recall that we evolve z by (zCSS), which is independent of η. If we define

$$w(t,x) := Q_{b(t)}^{(\eta)\sharp}(x) + z(t,x),$$

then w solves

$$\begin{aligned} &i\partial_s w^\flat - L_{w^\flat}^* \mathbf{D}_+^{(w^\flat)} w^\flat + ib\Lambda w^\flat - \eta\theta_\eta w^\flat \\ &= i\Big(\frac{\lambda_s}{\lambda} + b\Big)\Lambda z^\flat + (\gamma_s - \eta\theta_\eta) z^\flat + (b_s + b^2 + \eta^2)\tfrac{|y|^2}{4} Q_b^{(\eta)} \\ &\quad - V_{Q_b^{(\eta)} - Q_b} z^\flat + (V_{Q_b^{(\eta)} \to z^\flat} z^\flat - R_{Q_b^{(\eta)}, z^\flat}), \end{aligned}$$

where $V_{Q_b^{(\eta)} - Q_b}$ is an additional error induced from the external potential.

$$V_{Q_b^{(\eta)} - Q_b} := V_{Q_b^{(\eta)} \to z^\flat} - V_{Q_b \to z^\flat}. \tag{5.13}$$

Then the evolution of ϵ is given by

$$\begin{aligned} &i\partial_s \epsilon - \mathcal{L}_{w^\flat}\epsilon + ib\Lambda\epsilon - \eta\theta_\eta \epsilon \\ &= i\Big(\frac{\lambda_s}{\lambda} + b\Big)\Lambda(Q_b^{(\eta)} + \epsilon) + (\widetilde{\gamma}_s - \eta\theta_\eta) Q_b^{(\eta)} + (\gamma_s - \eta\theta_\eta)\epsilon \\ &\quad - (b_s + b^2 + \eta^2)\tfrac{|y|^2}{4} Q_b^{(\eta)} + \widetilde{R}_{Q_b^{(\eta)}, z^\flat} + V_{Q_b^{(\eta)} - Q_b} z^\flat + R_{u^\flat - w^\flat}. \end{aligned} \tag{5.14}$$

Applying the $\sharp$ operation, we get

$$\begin{aligned} i\partial_t \epsilon^\sharp - \mathcal{L}_w \epsilon^\sharp &= \frac{1}{\lambda^2} i\Big(\frac{\lambda_s}{\lambda} + b\Big)\Lambda Q_b^{(\eta)\sharp} + \frac{1}{\lambda^2}(\widetilde{\gamma}_s - \eta\theta_\eta) Q_b^{(\eta)\sharp} \\ &\quad - \frac{1}{\lambda^2}(b_s + b^2 + \eta^2)[\tfrac{|y|^2}{4} Q^{(\eta)}]_b^\sharp + \widetilde{R}_{Q_b^{(\eta)\sharp}, z} + V_{Q_b^{(\eta)\sharp} - Q_b^\sharp} z + R_{u-w}. \end{aligned} \tag{5.15}$$

5.3. Choice of modulation parameters

As we have three modulation parameters b, λ, and γ, we can impose three conditions to fix dynamical laws and $\epsilon(s,y)$. In order to guarantee the coercivity of the linearized operator, we spend two degrees of freedom for orthogonality conditions. Fix any smooth compactly supported real-valued functions $\mathcal{Z}_{\mathrm{re}}, \mathcal{Z}_{\mathrm{im}} : (0,\infty) \to \mathbb{R}$ such that

$$(\mathcal{Z}_{\mathrm{re}}, \Lambda Q)_r = (\mathcal{Z}_{\mathrm{im}}, Q)_r = 1 \neq 0. \tag{5.16}$$

We impose the orthogonality conditions as

$$(\epsilon, [\mathcal{Z}_{\mathrm{re}}]_b)_r = (\epsilon, [i\mathcal{Z}_{\mathrm{im}}]_b)_r = 0. \tag{5.17}$$

In view of Lemma 3.9, (5.17) implies coercivity of the linearized operator, provided that b is sufficiently small. A cleverer choice of the orthogonality conditions is not important in our analysis.

For the remaining one degree of freedom, we impose a dynamical law between b and λ as

$$2\Big(\frac{\lambda_s}{\lambda} + b\Big)b - (b_s + b^2 + \eta^2) = 0. \tag{5.18}$$

This is motivated to cancel out $\frac{|y|^2}{4} Q_b^{(\eta)}$ in the equation (5.14) of ϵ, since it has slow decay $|\frac{|y|^2}{4} Q_b| \sim r^{-m}$ as $r \to \infty$. Such a decay is not sufficient to close our bootstrap argument, particularly when m is small. Using $\Lambda Q_b^{(\eta)} = [\Lambda Q^{(\eta)}]_b - ib\frac{|y|^2}{2} Q_b^{(\eta)}$, we organize terms containing $\frac{|y|^2}{4} Q_b^{(\eta)}$ in (5.14) as

$$\begin{aligned} &i\Big(\frac{\lambda_s}{\lambda} + b\Big)\Lambda Q_b^{(\eta)} - (b_s + b^2 + \eta^2)\tfrac{|y|^2}{4} Q_b^{(\eta)} \\ &= \Big[2\Big(\frac{\lambda_s}{\lambda} + b\Big)b - (b_s + b^2 + \eta^2)\Big]\tfrac{|y|^2}{4} Q_b^{(\eta)} + i\Big(\frac{\lambda_s}{\lambda} + b\Big)[\Lambda Q^{(\eta)}]_b. \end{aligned}$$

By (5.18), we can delete the $\frac{|y|^2}{4} Q_b^{(\eta)}$-contribution in (5.14); we have

$$i\Big(\frac{\lambda_s}{\lambda} + b\Big)\Lambda Q_b^{(\eta)} - (b_s + b^2 + \eta^2)\tfrac{|y|^2}{4} Q_b^{(\eta)} = i\Big(\frac{\lambda_s}{\lambda} + b\Big)[\Lambda Q^{(\eta)}]_b.$$

This cancellation is heavily used in Sections 6.4 and 6.5. Note that (5.18) is consistent with

$$b(t) = -t \quad \text{and} \quad \lambda(t) = \langle t \rangle.$$

We now discuss on the existence of (λ, γ, b) satisfying (5.18) and orthogonality conditions (5.17). When $\eta = 0$, (5.18) is *explicitly solvable*; one sees that

$$\Big(\frac{b}{\lambda^2}\Big)_t = \frac{1}{\lambda^2}\Big(\frac{b}{\lambda^2}\Big)_s = \frac{b}{\lambda^4}\Big(\frac{b_s}{b} - 2\frac{\lambda_s}{\lambda}\Big) = \Big(\frac{b}{\lambda^2}\Big)^2. \tag{5.19}$$

This is satisfied by the choice

$$b(t) = |t|^{-1}\lambda^2(t) \tag{5.20}$$

for any time t. As we expect $\lambda^2(t) \sim |t|^2$ in our blow-up regime, (5.20) says that $|b(t)| \lesssim |t|$, which is small provided t is small. From this, one is able to use the implicit function theorem at each fixed time t.

When $\eta > 0$, we are not able to solve (5.18) to obtain a fixed time relation between b and λ, such as (5.20). Instead, we solve a system of ODEs for (b, λ, γ)

given by two orthogonality conditions and (5.18). The precise setup is as follows. We consider the initial data

$$u(0,x)=\frac{e^{i\gamma(0)}}{\lambda(0)}Q_{b(0)}^{(\eta)}\Big(\frac{x}{\lambda(0)}\Big)+z^{*}(x). \tag{5.21}$$

We evolve u by (CSS) and z by (zCSS). As we assume the decomposition

$$u(t,x)=\frac{e^{i\gamma(t)}}{\lambda(t)}[Q_{b(t)}^{(\eta)}+\epsilon](t,\frac{x}{\lambda(t)})+z(t,x),$$

ϵ is a function of u,z,λ,b,γ. Note that u and z are by now given, so ϵ is viewed as a function of λ,b,γ. The orthogonality conditions are viewed as first-order differential equations of (b,λ,γ), obtained by differentiating them in the t-variable and substituting (5.15). We also interpret (5.18) in the t-variable. Thus we have a system of ODEs

$$\begin{cases} \partial_t(\epsilon,[\mathcal{Z}_{\mathrm{re}}]_b)_r=0,\\ \partial_t(\epsilon,i[\mathcal{Z}_{\mathrm{im}}]_b)_r=0,\\ 2(\lambda\lambda_t+b)b-(\lambda^2 b_t+b^2+\eta^2)=0. \end{cases} \tag{5.22}$$

From (5.16), the system (5.22) is nondegenerate. In view of (6.27), the first and second lines of (5.22) essentially govern the laws of λ and γ. The third line governs the law of b. We then give the initial data at $t=0$ as $(b(0),\lambda(0),\gamma(0))=(0,\eta,0)$. Local existence and uniqueness are clear from Picard's theorem. Moreover, we can solve the system as long as the parameters b, $\log\lambda$, and γ do not diverge.

In the following sections, we will verify estimates of b,λ,γ, and ϵ via a bootstrap argument. The conclusion of the bootstrap procedure guarantees the existence of the decomposition.[2]

5.4. Reduction of Theorems 1.1 and 1.3 to the main bootstrap Lemma 5.3

We will construct a family of solutions $\{u^{(\eta)}\}_{\eta\in[0,\eta^*]}$ such that $u^{(0)}$ is a pseudo-conformal blow-up solutions. We first construct a sequence of solutions $\{u^{(\eta)}\}_{\eta\in(0,\eta^*]}$ on $[t_0^*,0]$ that approximately follow (4.12). Then by taking the limit $\eta\to 0$, we construct a pseudoconformal blow-up solution $u^{(0)}$ on $[t_0^*,0)$. A similar procedure was presented in Merle [**44**] and Merle-Raphaël-Szeftel [**50**].

Define

$$\gamma_{\mathrm{cor}}^{(\eta)}(t):=-\int_0^t \theta_{z\to Q_b^{(\eta)\sharp}}(t')dt', \tag{5.23}$$

where $\theta_{z\to Q_b^{(\eta)\sharp}}$ is as in (5.7) and (5.8). When $\eta=0$, we abbreviate it as $\gamma_{\mathrm{cor}}:=\gamma_{\mathrm{cor}}^{(\eta)}$.

PROPOSITION 5.2. *There exist $-1<t_0^*<0$ and $0<\eta^*<1$ such that for all $\eta\in(0,\eta^*]$ and all sufficiently small $0<\alpha^*<1$, we have the following property.*

[2] In the forward problems, when $u(t_0)=Q+\epsilon(t_0)$ is given, one has to do initial decomposition $u(t_0)=Q_{\mathcal{P}(t_0)}+\widetilde{\epsilon}(t_0)$ with some modulated profile $Q_{\mathcal{P}(t_0)}$ to guarantee that $\widetilde{\epsilon}$ satisfies some orthogonality conditions. For this purpose, one has to use the implicit function theorem at time $t=t_0$. In our case, the orthogonality condition at time $t=0$ is automatic as $\epsilon(0)=0$. This makes the argument simpler.

Let an m-equivariant profile z^* satisfy (H) *and $z(t,x)$ solve* (zCSS) *with the initial data $z(0,x) = z^*(x)$. Then the solution u to* (CSS) *with the initial data*

$$u^{(\eta)}(0,x) = \frac{1}{\eta} Q^{(\eta)}\Big(\frac{x}{\eta}\Big) + z^*(x)$$

is global-in-time and scatters. Moreover, $u^{(\eta)}$ admits a decomposition on time interval $[t_0^*, -t_0^*]$[3]

$$u^{(\eta)}(t,x) = \frac{e^{i\gamma(t)}}{\lambda(t)} Q^{(\eta)}_{b(t)}\Big(\frac{x}{\lambda(t)}\Big) + z(t,x) + \epsilon^\sharp(t,x)$$

with the estimates

$$\langle t\rangle^{-\frac14}\|\epsilon^\sharp\|_{L^2} + \|\epsilon^\sharp\|_{\dot H^1_m} + \Big|\frac{\lambda}{\langle t\rangle} - 1\Big| + \frac{|b+t|}{\lambda} + |\gamma - \gamma_\eta - \gamma^{(\eta)}_{\mathrm{cor}}| \lesssim \alpha^*(\langle t\rangle^{m+1} + \langle t\rangle^{\frac14}\eta^{\frac34}),$$

where γ_η is as in (4.12) *and $\langle t\rangle = (t^2+\eta^2)^{\frac12}$ is as in* (2.4).

PROOF OF PROPOSITION 5.2. Assuming the bootstrap lemma (Lemma 5.3) below, note that z is a global small solution (Section 5.1) with $z(0,x) = z^*(x)$. We will choose $\eta^* \in (0, \eta^{**}]$ later.

For $\eta \in (0,\eta^*)$, let $u^{(\eta)} : (T_-, T_+) \times \mathbb{R}^2 \to \mathbb{C}$ be the maximal lifespan solution to (CSS) with the initial data at time $t = 0$ as

$$u^{(\eta)}(0,x) = \frac{1}{\eta} Q^{(\eta)}\Big(\frac{x}{\eta}\Big) + z^*(x).$$

With the initial data

$$(b(0), \lambda(0), \gamma(0)) = (0, \eta, 0),$$

let $(b,\lambda,\gamma) : (\widetilde{T}_-, 0] \to \mathbb{R} \times \mathbb{R}_+ \times \mathbb{R}$ be the maximal lifespan solution to the system (5.22). Note that $T_- \le \widetilde{T}_- < 0$. Thus we have a decomposition of $u^{(\eta)}$ on $(\widetilde{T}_-, 0]$ satisfying

$$\begin{cases} u^{(\eta)}(t,x) = \dfrac{e^{i\gamma(t)}}{\lambda(t)} Q^{(\eta)}_{b(t)}\Big(\dfrac{x}{\lambda(t)}\Big) + z(t,x) + \epsilon^\sharp(t,x), \\ (\epsilon, [\mathcal{Z}_{\mathrm{re}}]_b)_r = (\epsilon, i[\mathcal{Z}_{\mathrm{im}}]_b)_r = 0, \\ 2\Big(\dfrac{\lambda_s}{\lambda} + b\Big) b - (b_s + b^2 + \eta^2) = 0. \end{cases} \tag{5.24}$$

We can now introduce our main bootstrap lemma.

LEMMA 5.3 (Main bootstrap). *There exists $t_0^* < 0$ such that for all sufficiently small $\eta > 0$ and $\alpha^* > 0$, we have the following property. Assume that the decomposition* (5.24) *of $u^{(\eta)}$ satisfies the weak bootstrap hypothesis*

$$\sup_{t\in[t_1,0]} \Big\{ \Big|\frac{\lambda}{\langle t\rangle} - 1\Big| + \frac{|b - |t||}{\lambda} \Big\} \le \frac12 \quad \text{and} \quad \sup_{t\in[t_1,0]} \Big\{ \frac{\|\epsilon\|_{\dot H^1_m}}{\lambda^{\frac32}} + \frac{\|\epsilon\|_{L^2}}{\lambda^{\frac12}} \Big\} \le 1 \tag{5.25}$$

[3]In fact, $\epsilon^\sharp$ and modulation parameters b, λ, and γ depend on η. We suppress the dependence on η to simplify the notations.

for some $t_1 \in [t_0^*, 0] \cap (\widetilde{T}_-, 0]$. *Then, the following strong conclusion holds for all* $t \in [t_1, 0]$*:*

$$\left\{ \begin{aligned} &\Big|\frac{\lambda}{\langle t\rangle} - 1\Big| + \frac{|b-|t||}{\lambda} + |\gamma - \gamma_\eta - \gamma_{\mathrm{cor}}^{(\eta)}| \lesssim \alpha^*(\lambda^{m+1} + \lambda^{\frac{1}{4}}\eta^{\frac{3}{4}}), \\ &\qquad\qquad \lambda^{\frac{3}{4}}\|\epsilon\|_{L^2} + \|\epsilon\|_{\dot{H}^1_m} \lesssim \alpha^*(\lambda^{m+2} + \lambda^{\frac{5}{4}}\eta^{\frac{3}{4}}), \end{aligned} \right. \tag{5.26}$$

where γ_η *as in* (4.12) *and* (4.10).

Let t_0^* be as in Lemma 5.3. Let η and α^* be sufficiently small to satisfy the hypothesis of Lemma 5.3. As $b(0) = 0$, $\lambda(0) = \eta$, and $\epsilon(0) = 0$, the bootstrap hypothesis (5.25) is satisfied in a neighborhood of the time 0. By a standard continuity argument, the bootstrap conclusion (5.26) is satisfied on the time interval $[t_0^*, 0] \cap (\widetilde{T}_-, 0]$. In particular, the modulation parameters b, λ, and γ do not blow up on $[t_0^*, 0] \cap (\widetilde{T}_-, 0]$ and hence $T_- \le \widetilde{T}_- < t_0^*$.

Finally, to see that $u^{(\eta)}$ scatters backward in time, observe that

$$\begin{aligned} &\Big\|u^{(\eta)}(t_0^*) - \frac{e^{i\gamma(t_0^*)}}{|t_0^*|} Q_{|t_0^*|}\Big(\frac{x}{|t_0^*|}\Big)\Big\|_{H^1} \\ &\le \|\epsilon^\sharp(t_0^*)\|_{H^1} + \|z(t_0^*)\|_{H^1} + \Big\|e^{i\gamma(t_0^*)}\Big\{\frac{1}{\lambda(t_0^*)}Q_{b(t_0^*)}^{(\eta)}\Big(\frac{x}{\lambda(t_0^*)}\Big) - \frac{1}{|t_0^*|}Q_{|t_0^*|}\Big(\frac{x}{|t_0^*|}\Big)\Big\}\Big\|_{H^1}. \end{aligned}$$

The first two terms are bounded by α^*. Since t_0^* is fixed, the last term goes to zero by shrinking α^* and η using (5.26) and (4.20). This determines the smallness of η^*. Hence we can apply the standard perturbation theory of scattering solutions (Proposition 2.9). Thus u scatters backward in time.

So far, we have discussed the backward evolution of $u^{(\eta)}$. To show the forward scattering of $u^{(\eta)}$, we use the time reversal symmetry $u(t,r) \mapsto \overline{u(-t,r)}$ of (CSS).[4] Now the initial data has changed into $v^{(\eta)}(0,r) := \frac{1}{\eta}Q^{(\eta)}(\frac{r}{\eta}) + \overline{z^*(r)}$. It is easy to check that $\overline{z^*(r)e^{im\theta}}$ satisfies (H) with the same α^*. Let $v^{(\eta)}(t,r)$ be the backward evolution of (CSS) with the initial data $v^{(\eta)}(0,r)$. All the above properties are still satisfied by $v^{(\eta)}(t,r)$ for $t<0$. Therefore, we can transfer them to the forward evolution $u^{(\eta)}(-t,r) = \overline{v^{(\eta)}(t,r)}$ for $t<0$. □

For each $\eta \in (0, \eta^*]$, let $u^{(\eta)}$ be the solution constructed in Proposition 5.2. In order to use a limiting argument, we need precompactness of the solutions $\{u^{(\eta)}\}_{\eta\in(0,\eta^*]}$.

LEMMA 5.4 (L^2 precompactness). *The set* $\{u^{(\eta)}(t_0^*)\}_{\eta\in(0,\eta^*]} \subset L^2_m$ *is precompact in* L^2_m.

PROOF. By the estimates in Proposition 5.2, the set $\{u^{(\eta)}(t_0^*)\}_{\eta\in(0,\eta^*]}$ is bounded in H^1_m. Thus it suffices to show that it is spatially localized, i.e.

$$\lim_{R\to\infty} \sup_{\eta\in(0,\eta^*]} \|u^{(\eta)}(t_0^*)\|_{L^2(r\ge R)} = 0.$$

To achieve this, we use local conservation laws (1.8) to obtain

$$\partial_t\Big(\int (1-\chi_R)|u^{(\eta)}|^2\Big) = -2\int (\partial_r\chi_R)\mathrm{Im}(\overline{u^{(\eta)}}\partial_r u^{(\eta)}),$$

[4]This indeed matches the time reversal symmetry mentioned in Introduction.

where χ_R is a smooth cutoff to the region $r \lesssim R$, i.e. $\chi_R(r) = \chi(\frac{r}{R})$ for a fixed smooth cutoff χ. Using the decomposition of $u^{(\eta)}$, we have

$$\Big|\int(\partial_r\chi_R)\mathrm{Im}(\overline{u^{(\eta)}(t)}\partial_r u^{(\eta)}(t))\Big| \lesssim \frac{1}{R}.$$

Since

$$\|u^{(\eta)}(0)\|_{L^2(r\gtrsim R)} \lesssim R^{-(m+1)} + \|z^*\|_{L^2(r\gtrsim R)},$$

an application of the fundamental theorem of calculus shows that

$$\int_{\mathbb{R}^2}(1-\chi_R)|u^{(\eta)}(t_0^*)|^2 \lesssim \sup_{t\in[t_0^*,0]}\|z(t)\|_{L^2(r\gtrsim R)}^2 + \frac{1}{R}.$$

Using the fact that $z:[t_0^*,0]\to L_m^2$ is a continuous path (and hence has the compact image in L_m^2), the conclusion follows. □

We now complete the proof of Theorem 1.1.

Proof of Theorem 1.1. Choose a sequence $\eta_n\to 0$ and let $u^{(n)} = u^{(\eta_n)}$ be the associated solution to Proposition 5.2. We extract a solution from taking limit of solutions $u^{(n)}$. From H_m^1 boundedness and L_m^2 pre-compactness (Lemma 5.4), there exists a subsequence (still denoted by $u^{(n)}$) such that $u^{(n)}(t_0^*)\rightharpoonup u(t_0^*)$ weakly in H_m^1 and $u^{(n)}(t_0^*)\to u(t_0^*)$ strongly in L_m^2. Interpolating these, $u^{(n)}(t_0^*)\to u(t_0^*)$ strongly in H_m^{1-}. Passing to a further subsequence, we may assume that $b^{(n)}(t_0^*)$, $\lambda^{(n)}(t_0^*)$, and $\gamma^{(n)}(t_0^*)$ converge to some limit $b(t_0^*)$, $\lambda(t_0^*)$, and $\gamma(t_0^*)$. Note that $\lambda(t_0^*)\approx b(t_0^*)\approx|t_0^*|$ and $\gamma(t_0^*)\approx -\frac{\pi}{2}(\frac{m+1}{m})$ by (5.26). Let $u:(T_-,T_+)\times\mathbb{R}^2\to\mathbb{C}$ be the maximal lifespan solution with the initial data $u(t_0^*)$ at time $t=t_0^*$.

We now show that $(T_-,T_+)=(-\infty,0)$ and u scatters backward in time. As $u(t_0^*)$ is close to $\frac{e^{i\gamma(t_0^*)}}{|t_0^*|}Q_{|t_0^*|}(\frac{x}{|t_0^*|})$ in L^2, we have $T_-=-\infty$ and u scatters backward in time. On the other hand, for all $t_0^*<t<\min\{T_+,0\}$, we have $u^{(n)}(t)\to u(t)$ strongly in H_m^{1-}. Since $\|u^{(n)}(t)\|_{H^{1-}}\sim|t^2+\eta_n^2|^{-(\frac{1}{2}-)}$, we conclude that $T_+=0$ by the H_m^{1-}-subcritical local well-posedness (Proposition 2.10).

So far, we only know that u is a H_m^1-solution; we do not know that u really inherits the decomposition estimates of $u^{(n)}$ due to the weak convergence. Here, we will show that we can transfer the H_m^1-bound of $(\epsilon^\sharp)^{(n)}$ to our limit solution. Let us evolve b, λ, and γ under (5.22) for $\eta=0$ with the initial data $b(t_0^*)$, $\lambda(t_0^*)$, and $\gamma(t_0^*)$ at time $t=t_0$. Since $u^{(n)}\to u$ in $C_{(-\infty,0),\mathrm{loc}}H_m^{1-}$ and $(b^{(n)}(t_0^*),\lambda^{(n)}(t_0^*),\gamma(t_0^*))\to(b(t_0^*),\lambda(t_0^*),\gamma(t_0^*))$, we see that $b(t)$, $\lambda(t)$, and $\gamma(t)$ indeed exist for all $t\in[t_0^*,0)$ and $(b^{(n)}(t),\lambda^{(n)}(t),\gamma^{(n)}(t))\to(b(t),\lambda(t),\gamma(t))$. For each fixed $t\in[t_0^*,0)$, we note that

$$\gamma(t)=\lim_{n\to\infty}\gamma^{(\eta_n)}(t)=\lim_{n\to\infty}\gamma_{\eta_n}(t)+O(\alpha^*|t|)=-(m+1)\frac{\pi}{2}+O(\alpha^*|t|), \tag{5.27}$$

where we estimated γ_{cor} by (5.23) and (6.6). We write

$$u(t,x)=\frac{1}{\lambda(t)}Q_{b(t)}\Big(\frac{x}{\lambda(t)}\Big)e^{i\gamma(t)}+z(t,x)+\epsilon^\sharp(t,x).$$

Thus $(\epsilon^\sharp)^{(n)}(t)\to\epsilon^\sharp(t)$ strongly in L_m^2. By the Fatou property, the H_m^1-bound of $\epsilon^\sharp$ is now transferred from that of $(\epsilon^\sharp)^{(n)}$. Therefore, we have shown that

$$|t|^{-\frac{1}{4}}\|\epsilon^\sharp\|_{L^2}+\|\epsilon^\sharp\|_{\dot H_m^1}+\Big|\frac{\lambda}{|t|}-1\Big|+\frac{|b+t|}{\lambda}+|\gamma+(m+1)\frac{\pi}{2}-\gamma_{\mathrm{cor}}|\lesssim\alpha^*|t|^{m+1}.$$

We now show (1.32). We first claim (5.20)

$$b(t) = |t|^{-1}\lambda^2(t).$$

To see this, we take a limit $\eta \to 0$ of (5.18) to get

$$2\Big(\frac{\lambda_s}{\lambda} + b\Big)b - (b_s + b^2) = 0.$$

Because of (5.19), the quantity

$$\frac{\lambda^2}{b} + t$$

is conserved. Since $\frac{\lambda^2}{b} + t \to 0$ as $t \to 0^-$, the claim follows. Therefore, we apply Lemma 7.1 to have

$$\begin{aligned}
&\|u(t,r) - \frac{1}{|t|}Q_{|t|}\Big(\frac{r}{|t|}\Big)e^{i(-(m+1)\frac{\pi}{2}+\gamma_{\mathrm{cor}})} - z(t,r)\|_{\dot H^1_m} \\
&\leq \|\epsilon^\sharp\|_{\dot H^1_m} + \frac{1}{|t|}\|\frac{|t|}{\lambda}Q_b\Big(\frac{|t|}{\lambda}r\Big)e^{i(\gamma+(m+1)\frac{\pi}{2}-\gamma_{\mathrm{cor}})} - Q_{|t|}(r)\|_{\dot H^1_m} \\
&\lesssim \|\epsilon^\sharp\|_{\dot H^1_m} + \frac{1}{|t|}\Big(\Big|\frac{\lambda}{|t|} - 1\Big| + |\gamma + (m+1)\frac{\pi}{2} - \gamma_{\mathrm{cor}}|\Big) \\
&\lesssim \alpha^*|t|^m.
\end{aligned}$$

One can similarly estimate

$$\begin{aligned}
&\|u(t,r) - \frac{1}{|t|}Q_{|t|}\Big(\frac{r}{|t|}\Big)e^{i(-(m+1)\frac{\pi}{2}+\gamma_{\mathrm{cor}})} - z(t,r)\|_{L^2} \\
&\lesssim \|\epsilon^\sharp\|_{L^2} + \Big(\Big|\frac{\lambda}{|t|} - 1\Big| + |\gamma + (m+1)\frac{\pi}{2} - \gamma_{\mathrm{cor}}|\Big) \\
&\lesssim \alpha^*|t|^{m+1}.
\end{aligned}$$

To complete the proof of Theorem 1.1, we apply the above procedure for the reversely rotated $e^{-i\frac{(m+1)\pi}{2}}z^*$ instead of z^*. We then apply the phase invariance $u(t,r) \mapsto e^{i\frac{(m+1)\pi}{2}}u(t,r)$ to get (1.31). □

Remark 5.5 (Alternative proof of Theorem 1.1). In fact, in order to construct a pseudoconformal blow-up solution, one can proceed all the above analysis only with $\eta = 0$. When $\eta = 0$, we cannot impose the initial data as in (5.21). Instead, one can consider a sequence of solutions $\{u^{(n)}\}_{n\in\mathbb{N}}$ with the initial data at $t = t_n$ as

$$u^{(n)}(t_n, r) = \frac{1}{|t_n|}Q_{|t_n|}\Big(\frac{r}{|t_n|}\Big) + z(t_n, r),$$

where $t_0^* < t_n < 0$ and $t_n \to 0^-$. One then takes the limit $n \to \infty$.

We defer the proof of Theorem 1.2 to Chapter 7. We now prove Theorem 1.3 and Corollary 1.4.

Proof of Theorem 1.3. The proof is almost immediate by combining Proposition 5.2, and Theorems 1.1 and 1.2. One caveat is that one should take into account the phase rotation in (5.27) as in the end of the proof of Theorem 1.1. Indeed, we apply Proposition 5.2 to obtain a family of solutions $\{u^{(\eta)}\}_{\eta\in(0,\eta^*]}$ with initial data

$$u^{(\eta)}(0,x) = \frac{1}{\eta}Q^{(\eta)}\Big(\frac{x}{\eta}\Big)e^{i(m+1)\frac{\pi}{2}} + z^*(x).$$

Let u be the solution constructed in Theorem 1.1. For any sequence $\eta_n \to 0$ in $(0, \eta^*]$, there exists a further sequence $\{\eta_{n_k}\}$ such that $u^{(\eta_{n_k})} \to \widetilde{u}$ in the $C_{(-\infty,0),\mathrm{loc}} H_m^{1-}$-topology. By the proof of Theorem 1.1, $\widetilde{u}$ satisfies (1.32). Applying Theorem 1.2, we should have $\widetilde{u} = u$. Therefore, $u^{(\eta)} \to u$ in the $C_{(-\infty,0),\mathrm{loc}} H_m^{1-}$-topology. The rest of the conclusions are contained in Proposition 5.2. □

Proof of Corollary 1.4. If $\eta > 0$, then $u^{(\eta)}$ is well-defined on $(-\infty, 0]$ and $\|u^{(\eta)}\|_{L^4_{(-\infty,0],x}} < \infty$. Therefore, $\|\mathcal{C}u^{(\eta)}\|_{L^4_{(0,\infty),x}} < \infty$ and hence $\mathcal{C}u^{(\eta)}$ scatters forward in time. If $\eta = 0$, we have

$$\|\mathcal{C}u - Q - \mathcal{C}z\|_{L^2} \lesssim \alpha^* |t|^{-1} \qquad \text{as } t \to +\infty.$$

Since z evolves under the $-(m+2)$-equivariant (CSS), $\mathcal{C}z$ scatters under the evolution $i\partial_t + \Delta_{-m-2}$. By the remark below, the proof is complete. □

Remark 5.6. A scattering solution under $i\partial_t + \Delta_{-m-2}$ also scatters under $i\partial_t + \Delta_m$. In view of $\Delta_{-m-2} = \Delta_m - \frac{4m+4}{r^2}$ and $m \neq 0$, the argument in [**4**] shows $\|\frac{4m+4}{r^2} u\|_{L^{4/3}_{t,x}} < \infty$ whenever u is a linear solution to $i\partial_t u + \Delta_{-m-2} u = 0$ with L^2-initial data.

Therefore, it only remains to prove Lemma 5.3 *and the conditional uniqueness Theorem* 1.2*. In Chapter* 6*, we prove Lemma* 5.3*; we assume our bootstrap hypothesis* (5.25) *and proceed to prove the bootstrap conclusion* (5.26)*. In Chapter* 7*, we prove Theorem* 1.2*.*

CHAPTER 6

Proof of bootstrap Lemma 5.3

The main ingredient of the proof is the Lyapunov/virial functional method as in Martel [**38**], Raphaël-Szeftel [**57**], and Merle-Raphaël-Szeftel [**50**]. Let u be a solution on $(t_0^{**}, t_0]$ satisfying the bootstrap hypothesis (5.25) on $[t_1, t_0] \subset (t_0^{**}, t_0]$. Note that u already has the decomposition (5.24) satisfying (5.17) and (5.18). We will discuss the details in Section 6.5.

Recall (5.14). We start with estimates of $\widetilde{R}_{Q_b^{(\eta)}, z^\flat}$ and $R_{u^\flat - w^\flat}$.

6.1. Estimates of $\widetilde{R}_{Q_b^{(\eta)}, z^\flat}$

In Section 5.1, we have extracted the strong interactions between $Q_b^{(\eta)}$ and $z^\flat$, which are $V_{Q_b^{(\eta)} \to z^\flat} z^\flat$ and $\theta_{z^\flat \to Q_b^{(\eta)}} Q_b^{(\eta)}$. The remaining part

$$\widetilde{R}_{Q_b^{(\eta)}, z^\flat} = R_{Q_b^{(\eta)}, z^\flat} - V_{Q_b^{(\eta)} \to z^\flat} z^\flat - \theta_{z^\flat \to Q_b^{(\eta)}} Q_b^{(\eta)}$$

will be estimated as an error. Note that this is independent of ϵ and hence eventually gives an error bound of ϵ.

In estimates of the size of $\widetilde{R}_{Q_b^{(\eta)}, z^\flat}$, the main strategy is to exploit decoupling of $Q_b^{(\eta)}$ and $z^\flat$ in their scales as in [**3**]. It is convenient to introduce functions $F_{n,\ell}$ and G_k on $(0, \infty)$ such that

$$F_{n,\ell} := \begin{cases} r^n & \text{if } r \leq 1, \\ r^{-\ell} & \text{if } r > 1, \end{cases} \tag{6.1}$$

and

$$G_k := \begin{cases} |\lambda r|^k & \text{if } r \leq \lambda^{-1}, \\ 1 & \text{if } r > \lambda^{-1}. \end{cases} \tag{6.2}$$

We note that $|Q_b^{(\eta)}| \lesssim F_{m,m+2}$ and $|z^\flat| \lesssim \alpha^* \lambda G_{m+2}$. As F and G lie in different scales, we have the following decoupling lemma:

LEMMA 6.1 (Estimates for decoupling at different scales). *Let $1 \leq p \leq \infty$; let $k, \ell, n \in \mathbb{R}$ be such that $\ell > \frac{2}{p}$ and $n + k + \frac{2}{p} > 0$. Then,*

$$\|F_{n,\ell} G_k\|_{L^p} \lesssim_{k,\ell,n,p} \lambda^{\min\{k, \ell - \frac{2}{p}\}}.$$

PROOF. The case of $p = \infty$ is immediate. For $1 \leq p < \infty$, we observe

$$\begin{aligned} \int_0^\infty F_{n,\ell}^p G_k^p r dr &\lesssim \int_0^1 |\lambda r|^{pk} r^{pn} r dr + \int_1^{\lambda^{-1}} |\lambda r|^{pk} r^{-p\ell} r dr + \int_{\lambda^{-1}}^\infty r^{-p\ell} r dr \\ &\lesssim_{k,\ell,n,p} \lambda^{pk} + \lambda^{p\ell - 2}. \end{aligned}$$

This completes the proof. ☐

We will use Lemma 6.1 to estimate $\widetilde{R}_{Q_b^{(\eta)},z^b}$. This is nothing but a rigorous justification of the heuristics in Section 5.1. For this purpose, we recall

$$L_u^*\mathbf{D}_+^{(u)}u = -(\partial_{rr}+\frac{1}{r}\partial_r)u + \Big(\frac{m+A_\theta[u]}{r}\Big)^2 u + A_0[u]u - |u|^2u.$$

Following the heuristics in Section 5.1, we reorganize $\widetilde{R}_{Q_b^{(\eta)},z^b}$ as

$$\begin{aligned}\widetilde{R}_{Q_b^{(\eta)},z^b} &= L^*_{Q_b^{(\eta)}+z^b}\mathbf{D}_+^{(Q_b^{(\eta)}+z^b)}(Q_b^{(\eta)}+z^b) - L^*_{Q_b^{(\eta)}}\mathbf{D}_+^{(Q_b^{(\eta)})}Q_b^{(\eta)} - L^*_{z^b}\mathbf{D}_+^{(z^b)}z^b \\ &\quad - V_{Q_b^{(\eta)}\to z^b}z^b - \theta_{z^b\to Q_b^{(\eta)}}Q_b^{(\eta)} \\ &= (6.3)+(6.4)+(6.5),\end{aligned}$$

where

$$(6.3)\qquad \begin{cases}\Big(\dfrac{m+A_\theta[Q^{(\eta)}+z^b]}{r}\Big)^2(Q^{(\eta)}+z^b) - \Big(\dfrac{m+A_\theta[Q^{(\eta)}]}{r}\Big)^2Q^{(\eta)} \\ \quad -\Big(\dfrac{m+A_\theta[Q^{(\eta)}](+\infty)+A_\theta[z^b]}{r}\Big)^2 z^b,\end{cases}$$

$$(6.4)\qquad \begin{cases}A_0[Q_b^{(\eta)}+z^b](Q_b^{(\eta)}+z^b) - A_0[Q_b^{(\eta)}]Q_b^{(\eta)} \\ +\Big(\displaystyle\int_r^\infty (m+A_\theta[Q^{(\eta)}](+\infty)+A_\theta[z^b])|z^b|^2\frac{dr'}{r'}\Big)(Q_b^{(\eta)}+z^b) \\ +\Big(\displaystyle\int_0^r (m+A_\theta[Q^{(\eta)}](+\infty)+A_\theta[z^b])|z^b|^2\frac{dr'}{r'}\Big)Q_b^{(\eta)},\end{cases}$$

$$(6.5)\qquad -|Q_b^{(\eta)}+z^b|^2(Q_b^{(\eta)}+z^b) + |Q_b^{(\eta)}|^2Q_b^{(\eta)} + |z^b|^2z^b.$$

Here, (6.3), (6.4), and (6.5) collect $(\frac{m+A_\theta[u]}{r})^2u$, $A_0[u]$, and $-|u|^2u$ parts of $L_u^*\mathbf{D}_+^{(u)}u$, respectively. The potential $V_{Q_b^{(\eta)}\to z^b}$ is distributed into (6.3) and (6.4), and $\theta_{z^b\to Q_b^{(\eta)}}$ contributes to (6.4). As explained in Section 5.1, in the introduction of $V_{Q_b^{(\eta)}\to z^b}$ and $\theta_{z^b\to Q_b^{(\eta)}}$, we have approximated $A_\theta[Q^{(\eta)}](r)$ by $A_\theta[Q^{(\eta)}](+\infty)$, and in A_0-term the integral $\int_r^\infty$ by $\int_0^\infty$. It turns out that the difference $A_\theta[Q]-A_\theta[Q](+\infty)$ exhibits fast decay at infinity and $\int_0^r$ integral enjoys degeneracy at the origin. So these differences can be covered by Lemma 6.1.

The following is the main result of this section.

LEMMA 6.2 (Estimates of $\widetilde{R}_{Q_b^{(\eta)},z^b}$). *We have*

$$|\theta_{z^b\to Q_b^{(\eta)}}| \lesssim (\alpha^*)^2\lambda^2, \tag{6.6}$$

$$\|\widetilde{R}_{Q_b^{(\eta)},z^b}\|_{H_m^1} \lesssim \alpha^*\lambda^{m+3}|\log\lambda|. \tag{6.7}$$

REMARK 6.3. The log factor of (6.7) comes from (6.15) below. This factor arises for all $m\geq 1$, as opposed to (4.1).

Setting aside the proof, we first collect some estimates of nonlinear terms in terms of F and G.

LEMMA 6.4 (Collection of various estimates). *We have the following.*

$$(6.8)\qquad \begin{cases}|Q_b^{(\eta)}| \lesssim F_{m,m+2}, \\ |\partial_rQ_b^{(\eta)}| + |\frac{1}{r}Q_b^{(\eta)}| \lesssim F_{m-1,m+3}+\lambda F_{m+1,m+1}.\end{cases}$$

$$\begin{cases} |A_\theta[Q^{(\eta)}]|_r^{+\infty}| =: |A_\theta[Q^{(\eta)}](+\infty) - A_\theta[Q^{(\eta)}](r)| \lesssim F_{0,2m+2}, \\ |\partial_r A_\theta[Q^{(\eta)}]| \lesssim F_{2m+1,2m+3}. \end{cases} \tag{6.9}$$

$$\begin{cases} \left| \int_r^\infty |Q_b^{(\eta)}|^2 \frac{dr'}{r'} \right| \lesssim F_{0,2m+4}, \\ \left| \partial_r \int_r^\infty |Q_b^{(\eta)}|^2 \frac{dr'}{r'} \right| \lesssim F_{2m-1,2m+5}. \end{cases} \tag{6.10}$$

$$\begin{cases} |z^b| \lesssim \alpha^* \lambda G_{m+2}, \\ |\partial_r z^b| + |\frac{1}{r} z^b| \lesssim \alpha^* \lambda^2 G_{m+1}. \end{cases} \tag{6.11}$$

$$\begin{cases} |A_\theta[z^b]| \lesssim (\alpha^*)^2 G_{2m+6}, \\ |\partial_r A_\theta[z^b]| \lesssim (\alpha^*)^2 \lambda G_{2m+5}. \end{cases} \tag{6.12}$$

$$\begin{cases} \int_0^r |z^b|^2 \frac{dr'}{r'} \lesssim (\alpha^*)^2 \lambda^2 G_{2m+4}, \\ \left| \partial_r \int_0^r |z^b|^2 \frac{dr'}{r'} \right| \lesssim (\alpha^*)^2 \lambda^3 G_{2m+3}. \end{cases} \tag{6.13}$$

$$\begin{cases} \left| \int_0^r \mathrm{Re}(\overline{Q_b^{(\eta)} z^b}) r' dr' \right| \lesssim \alpha^* \lambda^{m+1} F_{2m+2,0} G_2, \\ \left| \partial_r \int_0^r \mathrm{Re}(\overline{Q_b^{(\eta)} z^b}) r' dr' \right| \lesssim \alpha^* \lambda^{m+1} F_{2m+1,1} G_2. \end{cases} \tag{6.14}$$

$$\begin{cases} \left| \int_r^\infty \mathrm{Re}(\overline{Q_b^{(\eta)} z^b}) \frac{dr'}{r'} \right| \lesssim \alpha^* \lambda^{m+3} \cdot \begin{cases} 1 + |\log \lambda| & \textit{if } r \le 1, \\ 1 + |\log(\lambda r)| & \textit{if } 1 \le r \le \lambda^{-1} \\ 1 & \textit{if } r \ge \lambda^{-1}, \end{cases}, \\ \left| \partial_r \int_r^\infty \mathrm{Re}(\overline{Q_b^{(\eta)} z^b}) \frac{dr'}{r'} \right| \lesssim \alpha^* \lambda^{m+3}. \end{cases} \tag{6.15}$$

PROOF. The estimates (6.8)-(6.10) are immediate from (4.15) and explicit formula of Q. The estimate (6.11) follows from Lemma A.8. We turn to (6.12). In the region $r \le \lambda^{-1}$, (6.12) follows from (6.11). In the region $r \ge \lambda^{-1}$, we estimate $|A_\theta[z^b]| \lesssim \|z^b\|_{L^2}^2 \lesssim (\alpha^*)^2$ and $|\partial_r A_\theta[z^b]| \lesssim |r^{\frac{1}{2}} z^b|^2 \lesssim (\alpha^*)^2 \lambda$ by the Strauss inequality. The estimate (6.13) is similar to (6.12). Finally, the estimates (6.14) and (6.15) follow from (6.8) and (6.11). □

REMARK 6.5. In (6.8)-(6.15), the first estimate is adapted to L^2-estimate and the second is for $\dot{H}_m^1$-estimate. It is worth mentioning that in (6.8), we lose one decay by taking ∂_r but we compensate it by the factor λ. In (6.11)-(6.13), we lose one r-factor at origin by taking ∂_r but we again compensate it by the factor λ. For the remaining estimates, ∂_r-estimate is even better than the original estimate. This observation allows us to transfer L^2-estimate of $\widetilde{R}_{Q_b^{(\eta)},z^b}$ to H_m^1-estimate of $\widetilde{R}_{Q_b^{(\eta)},z^b}$.

PROOF OF LEMMA 6.2. Note that (6.6) follows from

$$|\theta_{z^b \to Q_b^{(\eta)}}| \lesssim \|r^{-1} z^b\|_{L^2}^2 \lesssim (\alpha^*)^2 \lambda^2.$$

We turn to (6.7). We estimate (6.3)-(6.5) term by term. We first start with (6.5). The mixed terms are estimated as

$$\begin{aligned}
&\|(Q_b^{(\eta)})^2 z^b\|_{L^2} \lesssim \alpha^* \lambda \|F_{m,m+2}^2 G_{m+2}\|_{L^2} \lesssim \alpha^* \lambda^{m+3}, \\
&\|Q_b^{(\eta)} (z^b)^2\|_{L^2} \lesssim (\alpha^* \lambda)^2 \|F_{m,m+2} G_{m+2}^2\|_{L^2} \lesssim (\alpha^*)^2 \lambda^{m+3}.
\end{aligned}$$

To get $\dot{H}_m^1$-estimate, we take ∂_r and $\frac{1}{r}$ to (6.5). Using the Leibniz rules, (6.8), and (6.11), the same estimate $\alpha^* \lambda^{m+3}$ follows.

We turn to (6.3). We expand

$$\begin{aligned}
(6.3) = {} & \frac{1}{r^2}\Big((m + A_\theta[Q^{(\eta)} + z^b])^2 - (m + A_\theta[Q^{(\eta)}] + A_\theta[z^b])^2\Big)(Q_b^{(\eta)} + z^b) \\
& + \frac{1}{r^2}\Big((m + A_\theta[Q^{(\eta)}] + A_\theta[z^b])^2 - (m + A_\theta[Q^{(\eta)}])^2\Big) Q_b^{(\eta)} \\
& + \frac{1}{r^2}\Big((m + A_\theta[Q^{(\eta)}] + A_\theta[z^b])^2 - (m + A_\theta[Q^{(\eta)}](+\infty) + A_\theta[z^b])^2\Big) z^b.
\end{aligned}$$

We view each line of the form

$$\frac{1}{r^2}(\alpha^2 - \beta^2)\varphi = (\alpha + \beta) \cdot \frac{1}{r^2}(\alpha - \beta)\varphi. \tag{6.16}$$

We discard $\alpha + \beta$ using $\|\alpha + \beta\|_{L^\infty} \lesssim 1$. For the first line, we estimate

$$\begin{aligned}
&\|\frac{1}{r^2}\Big(\int_0^r \mathrm{Re}(\overline{Q_b^{(\eta)}} z^b) r' dr'\Big) Q_b\|_{L^2} \lesssim \alpha^* \lambda^{m+1} \|F_{3m,m+4} G_2\|_{L^2} \lesssim \alpha^* \lambda^{m+3}, \\
&\|\frac{1}{r^2}\Big(\int_0^r \mathrm{Re}(\overline{Q_b^{(\eta)}} z^b) r' dr'\Big) z^b\|_{L^2} \lesssim (\alpha^*)^2 \lambda^{m+2} \|F_{2m,2} G_{m+4}\|_{L^2} \lesssim (\alpha^*)^2 \lambda^{m+3}.
\end{aligned}$$

For the second line, we estimate

$$\|\frac{1}{r^2} A_\theta[z^b] Q_b^{(\eta)}\|_{L^2} \lesssim (\alpha^*)^2 \|F_{m-2,m+4} G_{2m+6}\|_{L^2} \lesssim (\alpha^*)^2 \lambda^{m+3}.$$

For the last line, we estimate

$$\begin{aligned}
\|\frac{1}{r^2}(A_\theta[Q^{(\eta)}] - A_\theta[Q^{(\eta)}](+\infty)) z^b\|_{L^2} &= 2m \|\frac{1}{r^2}(A_\theta[Q^{(\eta)}]|_r^{+\infty}) z^b\|_{L^2} \\
&\lesssim \alpha^* \lambda \|F_{-2,2m+4} G_{m+2}\|_{L^2} \lesssim \alpha^* \lambda^{m+3}.
\end{aligned}$$

This concludes the L^2-estimate of (6.3). We turn to the $\dot{H}_m^1$-estimate. We take ∂_r and $\frac{1}{r}$ to (6.3). If ∂_r hits $\alpha + \beta$ part of (6.16), we use $\|\partial_r(\alpha + \beta)\|_{L^\infty} \lesssim 1$ followed from Lemma 6.4. If ∂_r hits $\frac{1}{r^2}$ of (6.16), then one can move one $\frac{1}{r}$ to φ and apply the derivative estimates of Lemma 6.4. If ∂_r and $\frac{1}{r}$ hits $(\alpha - \beta)\varphi$, then we apply the derivative estimates of Lemma 6.4. For example,

$$\begin{aligned}
&\|\partial_r\Big(\frac{1}{r^2} A_\theta[Q^{(\eta)}] z^b\Big)\|_{L^2} + \|\frac{1}{r}\Big(\frac{1}{r^2} A_\theta[Q^{(\eta)}] z^b\Big)\|_{L^2} \\
&\lesssim \|\frac{1}{r^2} A_\theta[Q^{(\eta)}] \frac{z^b}{r}\|_{L^2} + \|\frac{1}{r^2}(\partial_r A_\theta[Q^{(\eta)}]) z^b\|_{L^2} + \|\frac{1}{r^2} A_\theta[Q^{(\eta)}](\partial_r z^b)\|_{L^2} \\
&\lesssim \alpha^* \lambda^{m+3}.
\end{aligned}$$

The other terms can be treated similarly.

Finally, we consider (6.4).

$$\begin{aligned}
&A_0[Q_b^{(\eta)} + z^b](Q_b^{(\eta)} + z^b) - A_0[Q_b^{(\eta)}]Q_b^{(\eta)} \\
&+ \Big(\int_r^\infty (m + A_\theta[Q^{(\eta)}](+\infty) + A_\theta[z^b])|z^b|^2 \frac{dr'}{r'}\Big)(Q_b^{(\eta)} + z^b) \\
&+ \Big(\int_0^r (m + A_\theta[Q^{(\eta)}](+\infty) + A_\theta[z^b])|z^b|^2 \frac{dr'}{r'}\Big)Q_b^{(\eta)},
\end{aligned}$$

We expand

$$\begin{aligned}
(6.4) = &-2\Big(\int_r^\infty (m + A_\theta[Q_b^{(\eta)} + z^b])\mathrm{Re}(\overline{Q_b^{(\eta)}} z^b)\frac{dr'}{r'}\Big)(Q_b^{(\eta)} + z^b) \\
&+ \Big(\int_r^\infty \big(\int_0^r \mathrm{Re}(\overline{Q_b^{(\eta)}} z^b) r'dr'\big)(|Q_b^{(\eta)}|^2 + |z^b|^2)\frac{dr'}{r'}\Big)(Q_b^{(\eta)} + z^b) \\
&+ \Big(\int_r^\infty (A_\theta[Q^{(\eta)}](+\infty) - A_\theta[Q^{(\eta)}](r))|z^b|^2\frac{dr'}{r'}\Big)(Q_b^{(\eta)} + z^b) \\
&- \Big(\int_r^\infty (m + A_\theta[Q^{(\eta)}] + A_\theta[z^b])|Q_b^{(\eta)}|^2\frac{dr'}{r'}\Big)(Q_b^{(\eta)} + z^b) - A_0[Q_b^{(\eta)}]Q_b^{(\eta)} \\
&+ \Big(\int_0^r (m + A_\theta[Q^{(\eta)}](+\infty) + A_\theta[z^b])|z^b|^2\frac{dr'}{r'}\Big)Q_b^{(\eta)}.
\end{aligned}$$

We treat the above line by line. For the first line, we use (6.15) to estimate

$$\begin{aligned}
&\|\Big(\int_r^\infty (m + A_\theta[Q_b^{(\eta)} + z^b])\mathrm{Re}(\overline{Q_b^{(\eta)}} z^b)\tfrac{dr'}{r'}\Big)Q_b^{(\eta)}\|_{L^2} \\
&\lesssim \|m + A_\theta[Q_b^{(\eta)} + z^b]\|_{L^\infty}\|\Big(\int_r^\infty |Q_b^{(\eta)} z^b|\frac{dr'}{r'}\Big)\|_{L^\infty}\|Q_b^{(\eta)}\|_{L^2} \lesssim \alpha^*\lambda^{m+3}|\log\lambda|
\end{aligned} \tag{6.17}$$

and

$$\begin{aligned}
&\|\Big(\int_r^\infty (m + A_\theta[Q_b^{(\eta)} + z^b])\mathrm{Re}(\overline{Q_b^{(\eta)}} z^b)\tfrac{dr'}{r'}\Big)z^b\|_{L^2} \\
&\lesssim \|m + A_\theta[Q_b^{(\eta)} + z^b]\|_{L^\infty}\|\Big(\int_r^\infty |Q_b^{(\eta)} z^b|\frac{dr'}{r'}\Big)z^b\|_{L^2} \lesssim \alpha^*\lambda^{m+3}.
\end{aligned}$$

Note that (6.17) is the only term with the logarithmic factor. For sake of simplicity, we denote by ψ either $Q_b^{(\eta)}$ or z^b. For the second line, we estimate

$$\begin{aligned}
&\|\Big(\int_r^\infty \big(\int_0^r \mathrm{Re}(\overline{Q_b^{(\eta)}} z^b) r'dr'\big)|\psi|^2\frac{dr'}{r'}\Big)(Q_b^{(\eta)} + z^b)\|_{L^2} \\
&\lesssim \alpha^*\lambda^{m+1}\|\Big(\int_r^\infty F_{2m+2,0}G_2|\psi|^2\frac{dr'}{r'}\Big)(Q_b^{(\eta)} + z^b)\|_{L^2} \\
&\lesssim \alpha^*\lambda^{m+1}\cdot\begin{cases} \|F_{3m+2,m+2}G_2\|_{L^2}\|Q_b^{(\eta)} + z^b\|_{L^\infty} & \text{if } \psi = Q_b^{(\eta)}, \\ \|r^{-1}z^b\|_{L^2}^2\|Q_b^{(\eta)} + z^b\|_{L^2} & \text{if } \psi = z^b, \end{cases} \\
&\lesssim \alpha^*\lambda^{m+3}.
\end{aligned}$$

For the third line, we estimate

$$\begin{aligned}&\|(\int_r^\infty A_\theta[Q^{(\eta)}]|_r^{+\infty}|z^\flat|^2\tfrac{dr'}{r'})(Q_b^{(\eta)}+z^\flat)\|_{L^2}\\&\lesssim \|A_\theta[Q^{(\eta)}]|_r^{+\infty}|z^\flat|^2\|_{L^2}\|Q_b^{(\eta)}+z^\flat\|_{L^\infty}\\&\lesssim (\alpha^*)^2\lambda^2\|F_{0,2m+2}G_{2m+4}\|_{L^2}\lesssim(\alpha^*)^2\lambda^{2m+3}.\end{aligned}$$

For the fourth line, we estimate

$$\|(\int_r^\infty A_\theta[Q^{(\eta)}]|Q_b^{(\eta)}|^2\tfrac{dr'}{r'})z^\flat\|_{L^2}\lesssim \alpha^*\lambda\|F_{0,2m+4}G_{m+2}\|_{L^2}\lesssim\alpha^*\lambda^{m+3}$$

and

$$\begin{aligned}\|(\int_r^\infty A_\theta[z^\flat]|Q_b^{(\eta)}|^2\tfrac{dr'}{r'})(Q_b^{(\eta)}+z^\flat)\|_{L^2}&\lesssim \|A_\theta[z^\flat]|Q_b^{(\eta)}|^2\|_{L^2}\|Q_b^{(\eta)}+z^\flat\|_{L^\infty}\\&\lesssim(\alpha^*)^2\|G_{2m+6}F_{2m,2m+4}\|_{L^2}\lesssim(\alpha^*)^2\lambda^{2m+3}.\end{aligned}$$

For the last line, we estimate

$$\begin{aligned}&\|(\int_0^r(m+A_\theta[Q^{(\eta)}](+\infty)+A_\theta[z^\flat])|z^\flat|^2\tfrac{dr'}{r'})Q_b^{(\eta)}\|_{L^2}\\&\lesssim\|m+A_\theta[Q^{(\eta)}](+\infty)+A_\theta[z^\flat]\|_{L^\infty}\|(\int_0^r|z^\flat|^2\frac{dr'}{r'})Q_b^{(\eta)}\|_{L^2}\\&\lesssim(\alpha^*)^2\lambda^2\|F_{m,m+2}G_{2m+4}\|_{L^2}\lesssim(\alpha^*)^2\lambda^{m+3}.\end{aligned}$$

In order to get $\dot H_m^1$-estimate, use the algebra

$$\partial_r\Big(\int_r^\infty f\frac{dr'}{r'}\Big)=\int_r^\infty\Big(-(\partial_r f)+\frac{f}{r'}\Big)\frac{dr'}{r'}$$

for the outermost integral $\int_r^\infty$. One can then apply the derivative estimates of Lemma 6.4. This completes the proof. □

Indeed, (6.7) will be used in various places of the following sections. In most cases (6.7) suffices, but in Section 6.5 we need to improve some estimate on $\widetilde{R}_{Q_b^{(\eta)},z^\flat}$ by the $|\log\lambda|$ factor. Inspecting the previous proof, one observes that there is only one term (6.17) yielding the factor $|\log\lambda|$. In the following lemma, we will separate (6.17) and show the improved estimate (6.18) below. This is required to have the uniqueness of the pseudoconformal blow-up solutions constructed in Theorem 1.1. In fact, without the logarithmic improvement of the estimate, the error estimate of the energy identity (Lemma 6.20 and (6.39)) and hence those of Theorem 1.1 become worse logarithmically. This will not meet the hypothesis a priori bound of Theorem 1.2 when $m=1$.

LEMMA 6.6 (Logarithmic improvement). *Let*

$$R_2:=-2(\int_r^\infty(m+A_\theta[Q_b^{(\eta)}+z^\flat])\mathrm{Re}(\overline{Q_b^{(\eta)}}z^\flat)\tfrac{dr'}{r'})Q^{(\eta)}.$$

Then the decomposition

$$\widetilde{R}_{Q_b^{(\eta)},z^\flat}=:R_1+[R_2]_b$$

satisfies

$$\|R_1\|_{H_m^1}\lesssim\alpha^*\lambda^{m+3}$$

and

$$\|L_Q(iR_2)\|_{L^2} + \|r\partial_r L_Q(iR_2)\|_{L^2} \lesssim \alpha^* \lambda^{m+3}, \quad \|rR_2\|_{L^2} + \|r^2\partial_r R_2\|_{L^2} \lesssim \alpha^* \lambda^{m+3}|\log \lambda|. \tag{6.18}$$

PROOF. By the proof of (6.7), $\|R_1\|_{H^1_m} \lesssim \alpha^* \lambda^{m+3}$ is done. To prove estimates on R_2, let

$$f := -2\int_r^\infty (m + A_\theta[Q_b^{(\eta)} + z^b])\mathrm{Re}(\overline{Q_b^{(\eta)}} z^b)\frac{dr'}{r'}$$

so that $R_2 = fQ^{(\eta)}$. We compute using (4.8) that

$$\begin{aligned} L_Q(f \cdot iQ^{(\eta)}) &= (\partial_r f)\cdot iQ^{(\eta)} + f \cdot i\mathbf{D}_+^{(Q)} Q^{(\eta)} \\ &= (\partial_r f)\cdot iQ^{(\eta)} + f \cdot \tfrac{1}{r}(A_\theta[Q^{(\eta)}] - A_\theta[Q])iQ^{(\eta)} - f\cdot i\eta\tfrac{r}{2}Q^{(\eta)}. \end{aligned}$$

In view of (6.15) and $\eta \lesssim \lambda$, we have

$$\|\partial_r f\|_{L^\infty} + \|f(A_\theta[Q^{(\eta)}] - A_\theta[Q])\|_{L^\infty} + \|\eta f\|_{L^\infty} \lesssim \alpha^* \lambda^{m+3},$$

Therefore,

$$\|L_Q(f\cdot iQ^{(\eta)})\|_{L^2} + \|r\partial_r L_Q(f\cdot iQ^{(\eta)})\|_{L^2} \lesssim \alpha^*\lambda^{m+3}.$$

The remaining estimates follow from

$$\begin{aligned} \|f\|_{L^\infty} + \|r\partial_r f\|_{L^\infty} &\lesssim \alpha^*\lambda^{m+3}|\log\lambda|, \\ \|rQ^{(\eta)}\|_{L^2} + \|r^2\partial_r Q^{(\eta)}\|_{L^2} &\lesssim 1. \end{aligned}$$

□

We conclude this section with an estimate of $\widetilde{V}_{Q_b^{(\eta)}-Q_b}$ (recall (5.13)).

LEMMA 6.7 (Estimate of $\widetilde{V}_{Q_b^{(\eta)}-Q_b}$). *We have*

$$\|\widetilde{V}_{Q_b^{(\eta)}-Q_b} z^b\|_{H^1_m} \lesssim \alpha^*\lambda^2\eta. \tag{6.19}$$

PROOF. Recall from (4.17) that

$$|A_\theta[Q_b^{(\eta)}](+\infty) - A_\theta[Q_b](+\infty)| \lesssim \eta.$$

By scaling reasons, we have

$$\|\widetilde{V}_{Q_b^{(\eta)}-Q_b} z^b\|_{L^2} \lesssim \alpha^*\lambda^2\eta.$$

To show the derivative estimate, we observe

$$\begin{aligned} \|\widetilde{V}_{Q_b^{(\eta)}-Q_b} z^b\|_{\dot{H}^1_m} &\lesssim \sum_{T\in\{\partial_r, r^{-1}\}} \|\widetilde{V}_{Q_b^{(\eta)}-Q_b} T z^b\|_{L^2} + \|(|Q^{(\eta)}|^2 - |Q|^2)r^{-2}z^b\|_{L^2} \\ &\lesssim \alpha^*\lambda^3\eta + \alpha^*\lambda^2\||Q^{(\eta)} - Q|\cdot Q\|_{L^\infty} \\ &\lesssim \alpha^*\lambda^2\eta, \end{aligned}$$

where in the last inequality we used (4.16) and (4.15). This completes the proof of (6.19). □

6.2. Estimates of $R_{u^\flat - w^\flat}$, $\mathcal{L}_{w^\flat} - \mathcal{L}_{Q_b^{(\eta)}}$, and $\mathcal{L}_{Q_b^{(\eta)}} - \mathcal{L}_Q$

We will estimate each error terms with help of duality. In this section, we mainly rely on Lemma 2.2.

LEMMA 6.8 (Estimates of $R_{u^\flat - w^\flat}$). *We have*

$$\|R_{u^\flat - w^\flat}\|_{L^2} \lesssim \|\epsilon\|_{\dot{H}^1_m}^2, \tag{6.20}$$

$$\|(1+r)^{-1} R_{u^\flat - w^\flat}\|_{L^1} \lesssim (\|\epsilon\|_{L^2} + \|\epsilon\|_{\dot{H}^1_m})\|\epsilon\|_{\dot{H}^1_m}. \tag{6.21}$$

REMARK 6.9. The second estimate (6.21) is only used in case of $m = 1$, where we estimate $(R_{u^\flat - w^\flat}, |y|^2 Q_b^{(\eta)})_r$ in Section 6.5. The estimate (6.21) is not sharp, but this suffices to close our bootstrap.

PROOF. We show the first estimate (6.20). Note that

$$R_{u-w} = \sum_{\substack{\psi_1,\psi_2,\psi_3 \in \{Q_b^{(\eta)}, z^\flat, \epsilon\}: \\ \#\{i:\psi_i = \epsilon\} \geq 2}} [\mathcal{N}_{3,0} + \mathcal{N}_{3,1} + \mathcal{N}_{3,2}] + \sum_{\substack{\psi_1,\dots,\psi_5 \in \{Q_b^{(\eta)}, z^\flat, \epsilon\}: \\ \#\{i:\psi_i = \epsilon\} \geq 2}} [\mathcal{N}_{5,1} + \mathcal{N}_{5,2}].$$

We will use

$$\|\psi\|_{L^2} \lesssim 1 \qquad \forall \psi \in \{Q_b^{(\eta)}, z^\flat, \epsilon\}.$$

Now the estimate (6.20) follows by distributing two $\dot{H}^1_m$-norms to two ϵ's using (2.27).

To show the second estimate (6.21), we use

$$\|(1+r)^{-1} R_{u^\flat - w^\flat}\|_{L^1} \lesssim \|R_{u^\flat - w^\flat}\|_{L^{\frac{4}{3}}}.$$

We then apply Lemma 2.6. Here, we put $\dot{H}^1_m$-norm for one ϵ. For the remaining arguments, we put L^2 or L^4 norms as provided in Lemma 2.4. We then use $\|\epsilon\|_{L^2} + \|\epsilon\|_{L^4} \lesssim \|\epsilon\|_{L^2} + \|\epsilon\|_{\dot{H}^1_m}$. □

We turn to estimate the difference $\mathcal{L}_{w^\flat} - \mathcal{L}_{Q_b}$ and $\mathcal{L}_{Q_b} - \mathcal{L}_Q$. This will allow us to replace the linearized operator $\mathcal{L}_{w^\flat}$ by $\mathcal{L}_{Q_b}$ or $\mathcal{L}_Q$.

LEMMA 6.10 (Estimates of the linearized operators). *The following hold:*[1]

$$\|\mathcal{L}_{w^\flat} - \mathcal{L}_{Q_b^{(\eta)}}\|_{\dot{H}^1_m \to L^2} \lesssim \alpha^* \lambda, \tag{6.22}$$

$$\|\mathcal{L}_{w^\flat} - \mathcal{L}_{Q_b^{(\eta)}}\|_{\dot{H}^1_m \to (\dot{H}^1_m)^*} \lesssim \alpha^*, \tag{6.23}$$

$$\|\mathcal{L}_{Q_b^{(\eta)}} - \mathcal{L}_Q\|_{\dot{H}^1_m \to (\dot{H}^1_m)^*} \lesssim \lambda |\log \lambda|^{\frac{1}{2}}. \tag{6.24}$$

PROOF. We first consider the estimate for $\mathcal{L}_{w^\flat} - \mathcal{L}_{Q_b^{(\eta)}}$. Observe that

$$\begin{aligned}(\psi, \mathcal{L}_{w^\flat}\varphi - \mathcal{L}_{Q_b^{(\eta)}}\varphi)_r = &\sum_{\substack{\psi_1,\psi_2,\psi_3 \in \{Q_b^{(\eta)}, z^\flat, \varphi\}: \\ \#\{i:\psi_i = \varphi\} = 1 \\ \#\{i:\psi_i = z^\flat\} \geq 1}} (\psi, [\mathcal{N}_{3,0} + \mathcal{N}_{3,1} + \mathcal{N}_{3,2}](\psi_1,\psi_2,\psi_3))_r \\ &+ \sum_{\substack{\psi_1,\dots,\psi_5 \in \{Q_b^{(\eta)}, z^\flat, \varphi\}: \\ \#\{i:\psi_i = \varphi\} = 1 \\ \#\{i:\psi_i = z^\flat\} \geq 1}} (\psi, [\mathcal{N}_{5,1} + \mathcal{N}_{5,2}](\psi_1,\psi_2,\psi_3,\psi_4,\psi_5))_r.\end{aligned}$$

[1] In case of $m \geq 2$, we can replace $\lambda |\log \lambda|^{\frac{1}{2}}$ by λ.

We apply Lemma 2.2. If we put $\dot{H}^1_m$-norms to φ and $z^\flat$, put L^2-norms for remaining arguments, and use $\|z^\flat\|_{\dot{H}^1_m} \lesssim \alpha^*\lambda$, then $\dot{H}^1_m \to L^2$ estimate follows. If we put $\dot{H}^1_m$-norms to φ and ψ, put L^2-norms for remaining arguments, and use $\|z^\flat\|_{L^2} \lesssim \alpha^*$, then $\dot{H}^1_m \to (\dot{H}^1_m)^*$ estimate follows.

For $\mathcal{L}_{Q_b^{(\eta)}} - \mathcal{L}_Q$, the idea is exactly same with $\dot{H}^1_m \to (\dot{H}^1_m)^*$ estimate for $\mathcal{L}_{w^\flat} - \mathcal{L}_{Q_b^{(\eta)}}$. Use (4.1) and (4.20). Note that $b, \eta \lesssim \lambda$. We omit the details. $\square$

6.3. Modulation estimates

Applying (5.18) to the ϵ-equations (5.14) and (5.15), we write

$$\begin{aligned} i\partial_s\epsilon - \mathcal{L}_{w^\flat}\epsilon + ib\Lambda\epsilon - \eta\theta_\eta\epsilon &\\ = i\Big(\frac{\lambda_s}{\lambda} + b\Big)([\Lambda Q^{(\eta)}]_b + \Lambda\epsilon) &+ (\widetilde{\gamma}_s - \eta\theta_\eta)Q_b^{(\eta)} + (\gamma_s - \eta\theta_\eta)\epsilon \\ + \widetilde{R}_{Q_b^{(\eta)}, z^\flat} &+ V_{Q_b^{(\eta)} - Q_b} z^\flat + R_{u^\flat - w^\flat} \end{aligned} \tag{6.25}$$

and

$$\begin{aligned} i\partial_t\epsilon^\sharp - \mathcal{L}_w\epsilon^\sharp = \frac{1}{\lambda^2} i\Big(\frac{\lambda_s}{\lambda} + b\Big)[\Lambda Q^{(\eta)}]_b^\sharp + \frac{1}{\lambda^2}(\widetilde{\gamma}_s - \eta\theta_\eta)Q_b^{(\eta)\sharp} \\ + \widetilde{R}_{Q_b^{(\eta)\sharp}, z} + V_{Q_b^{(\eta)\sharp} - Q_b^\sharp} z + R_{u-w}. \end{aligned} \tag{6.26}$$

(6.25) is the equation of ϵ we use in what follows.

The main goal of this section is to obtain modulation estimates. Modulation estimates will be obtained from differentiating the orthogonality conditions.

LEMMA 6.11 (Computation of $\partial_s(\epsilon, \psi_b)_r$). *We have*

$$\begin{aligned} \partial_s(\epsilon, \psi_b)_r &= (\epsilon, [\mathcal{L}_{Q^{(\eta)}} i\psi + \eta\theta_\eta i\psi + \eta^2 \tfrac{i|y|^2\psi}{4}]_b)_r \\ &\quad + \Big(\frac{\lambda_s}{\lambda} + b\Big)(\Lambda Q^{(\eta)}, \psi)_r + (\widetilde{\gamma}_s - \eta\theta_\eta)(Q^{(\eta)}, i\psi)_r \\ &\quad - \Big(\frac{\lambda_s}{\lambda} + b\Big)(\epsilon, [\Lambda\psi]_b)_r + (\widetilde{\gamma}_s - \eta\theta_\eta)(\epsilon, i\psi_b)_r \\ &\quad - \theta_{z^\flat \to Q_b^{(\eta)}}(\epsilon, i\psi_b)_r + ((\mathcal{L}_{w^\flat} - \mathcal{L}_{Q_b^{(\eta)}})\epsilon, i\psi_b)_r + (\widetilde{R}_{Q_b^{(\eta)}, z^\flat}, i\psi_b)_r \\ &\quad + (V_{Q_b^{(\eta)} - Q_b} z^\flat, i\psi_b)_r + (R_{u^\flat - w^\flat}, i\psi_b)_r. \end{aligned} \tag{6.27}$$

PROOF. We compute

$$\begin{aligned} \partial_s(\epsilon, \psi_b)_r &= (i\partial_s\epsilon, i\psi_b)_r + b_s(\epsilon, \partial_b\psi_b)_r \\ &= (\mathcal{L}_{Q_b^{(\eta)}}\epsilon - ib\Lambda\epsilon + \eta\theta_\eta\epsilon, i\psi_b)_r \\ &\quad + \Big(\frac{\lambda_s}{\lambda} + b\Big)(\Lambda Q^{(\eta)}, \psi)_r + (\widetilde{\gamma}_s - \eta\theta_\eta)(Q^{(\eta)}, i\psi)_r \\ &\quad + \Big(\frac{\lambda_s}{\lambda} + b\Big)(\Lambda\epsilon, \psi_b)_r + (\gamma_s - \eta\theta_\eta)(\epsilon, i\psi_b)_r + b_s(\epsilon, \partial_b\psi_b)_r \\ &\quad + ((\mathcal{L}_{w^\flat} - \mathcal{L}_{Q_b^{(\eta)}})\epsilon, i\psi_b)_r + (\widetilde{R}_{Q_b^{(\eta)}, z^\flat}, i\psi_b)_r \\ &\quad + (V_{Q_b^{(\eta)} - Q_b} z^\flat, i\psi_b)_r + (R_{u^\flat - w^\flat}, i\psi_b)_r. \end{aligned}$$

We then use the self-adjointness of $\mathcal{L}_{Q_b^{(\eta)}}$, anti-self-adjointness of Λ, and (5.12) to get

$$\begin{aligned}\partial_s(\epsilon,\psi_b)_r &= (\epsilon, \mathcal{L}_{Q_b^{(\eta)}} i\psi_b + b\Lambda\psi_b + \eta\theta_\eta i\psi_b)_r \\ &\quad + \Big(\frac{\lambda_s}{\lambda}+b\Big)(\Lambda Q^{(\eta)},\psi)_r + (\widetilde{\gamma}_s - \eta\theta_\eta)(Q^{(\eta)}, i\psi)_r \\ &\quad - \Big(\frac{\lambda_s}{\lambda}+b\Big)(\epsilon,\Lambda\psi_b)_r + (\widetilde{\gamma}_s - \eta\theta_\eta)(\epsilon, i\psi_b)_r + b_s(\epsilon,\partial_b\psi_b)_r \\ &\quad - \theta_{z^b\to Q_b^{(\eta)}}(\epsilon, i\psi_b)_r + ((\mathcal{L}_{w^b} - \mathcal{L}_{Q_b^{(\eta)}})\epsilon, i\psi_b)_r + (\widetilde{R}_{Q_b^{(\eta)},z^b}, i\psi_b)_r \\ &\quad + (V_{Q_b^{(\eta)}-Q_b} z^b, i\psi_b)_r + (R_{u^b-w^b}, i\psi_b)_r.\end{aligned}$$

Observe by (5.18)

$$\begin{aligned}&-\Big(\frac{\lambda_s}{\lambda}+b\Big)(\epsilon,\Lambda\psi_b)_r + b_s(\epsilon,\partial_b\psi_b)_r \\ &= -\Big(\frac{\lambda_s}{\lambda}+b\Big)(\epsilon,[\Lambda\psi]_b)_r - (b^2+\eta^2)(\epsilon,\partial_b\psi_b)_r\end{aligned}$$

and by (4.3)

$$\mathcal{L}_{Q_b^{(\eta)}} i\psi_b + b\Lambda\psi_b - b^2\partial_b\psi_b = [\mathcal{L}_{Q^{(\eta)}} i\psi]_b.$$

Substituting the above two displays, we complete the proof. □

LEMMA 6.12 (Modulation estimates). *We have*

$$\Big|\frac{\lambda_s}{\lambda}+b\Big| + |\widetilde{\gamma}_s - \eta\theta^{(\eta)}| \lesssim \alpha^*\lambda^2(\lambda^{m+1}|\log\lambda| + \eta) + \|\epsilon\|_{\dot{H}^1_m}, \tag{6.28}$$

$$|b_s + b^2 + \eta^2| \lesssim \alpha^*\lambda^3(\lambda^{m+1}|\log\lambda| + \eta) + \lambda\|\epsilon\|_{\dot{H}^1_m}. \tag{6.29}$$

In particular, we have crude estimaes

$$\Big|\frac{\lambda_s}{\lambda}\Big| + |\gamma_s| \lesssim \lambda \quad \textit{and} \quad |b_s| \lesssim \lambda^2, \tag{6.30}$$

and positivity of b:

$$b \geq 0. \tag{6.31}$$

Moreover, we have a degeneracy estimate

$$(\epsilon, Q_b^{(\eta)})_r \lesssim \alpha^*\lambda(\lambda^{m+1}|\log\lambda| + \eta) + (\alpha^*+\lambda)\mathcal{T}^{(\eta,\frac{5}{4})}_{\dot{H}^1_m}[\epsilon]. \tag{6.32}$$

REMARK 6.13 (Refined modulation estimates). In the proof of Lemma 6.12, we use a crude estimate

$$|(\epsilon, [\mathcal{L}_{Q^{(\eta)}} i\psi + \eta\theta_\eta i\psi + \eta^2 \tfrac{i|y|^2\psi}{4}]_b)_r| \lesssim \|\epsilon\|_{\dot{H}^1_m} \tag{6.33}$$

for $\psi \in \{\mathcal{Z}_{\mathrm{re}}, i\mathcal{Z}_{\mathrm{im}}\}$, which is merely a consequence of Cauchy-Schwarz. Thus this estimate holds for any generic $\mathcal{Z}_{\mathrm{re}}$ and $\mathcal{Z}_{\mathrm{im}}$. However, if one cleverly chooses $\mathcal{Z}_{\mathrm{re}}$ and $\mathcal{Z}_{\mathrm{im}}$ adapted to the generalized nullspace of the linearized operator $\mathcal{L}_Q$, then one can indeed have better estimate. For example, if $\eta = 0$, $\mathcal{Z}_{\mathrm{re}} = \chi_M|y|^2Q$, and $\mathcal{Z}_{\mathrm{im}} = \chi_M\rho$, then one can obtain a maximal function version of

$$\Big|\frac{\lambda_s}{\lambda}+b\Big| + |\widetilde{\gamma}_s| \lesssim o_{M\to\infty}(1)\|L_Q\epsilon\|_{L^2} + \alpha^*\lambda^{m+2}|\log\lambda| + (\alpha^* + C(M)\lambda)\|\epsilon\|_{\dot{H}^1_m}.$$

If we look at coefficients of $\|L_Q\epsilon\|_{L^2}$ and $\|\epsilon\|_{\dot H^1_m}$, this can be viewed as an improved modulation estimate. However, our crude modulation estimate (6.28) still suffices to close bootstrap.

REMARK 6.14. Notice that (6.32) has a small coefficient $\alpha^* + \lambda$ of ϵ. This is better than a crude estimate $|(\epsilon, Q_b^{(\eta)})_r| \lesssim \|\epsilon\|_{\dot H^1_m}$. Such a gain arises because $iQ^{(\eta)}$ lies in the kernel of the operator $\mathcal{L}_{Q^{(\eta)}} + \eta\theta_\eta + \eta^2\frac{|y|^2}{4}$; see (4.24). This gain is essential to close our bootstrap procedure.

PROOF. Note that the bound (6.29) follows from (6.28) and (5.18). The crude estimate (6.30) follows from (6.28) and (6.29). The positivity (6.31) follows from $b(t=0)=0$, $b^2+\eta^2 \sim \lambda^2$, and (6.29) combined with the weak bootstrap hypothesis $\|\epsilon\|_{\dot H^1_m} \le \lambda^{\frac{3}{2}}$.

The modulation estimate (6.28) is obtained by differentiating the orthogonality conditions in the s-variable. Namely, we substitute $\psi \in \{\mathcal{Z}_{\mathrm{re}}, i\mathcal{Z}_{\mathrm{im}}\}$ into (6.27) and use $(\epsilon, \psi_b)_r = 0$.

The fourth and fifth lines of the RHS of (6.27) are treated as error. Indeed, we use (6.6), (6.22), (6.20), (6.19), and (6.7) to get

$$
\begin{cases}
|\theta_{z^\flat \to Q_b^{(\eta)}}(\epsilon, i\psi_b)_r| \lesssim (\alpha^*)^2\lambda^2\|\epsilon\|_{\dot H^1_m}, \\
|(\mathcal{L}_{w^\flat}\epsilon - \mathcal{L}_{Q_b^{(\eta)}}\epsilon, i\psi_b)_r| \lesssim \alpha^*\lambda\|\epsilon\|_{\dot H^1_m}, \\
|(R_{u^\flat - w^\flat}, i\psi_b)_r| \lesssim \|\epsilon\|_{\dot H^1_m}^2, \\
|(V_{Q_b^{(\eta)} - Q_b} z^\flat, i\psi_b)_r| \lesssim \alpha^*\lambda^2\eta. \\
|(\widetilde{R}_{Q_b^{(\eta)}, z^\flat}, i\psi_b)_r| \lesssim \alpha^*\lambda^{m+3}|\log\lambda|.
\end{cases}
\tag{6.34}
$$

This holds for $\psi \in \{\mathcal{Z}_{\mathrm{re}}, i\mathcal{Z}_{\mathrm{im}}, Q_b^{(\eta)}\}$. Therefore, the fourth and fifth lines of the RHS of (6.27) can be ignored from now on.

To extract $\frac{\lambda_s}{\lambda} + b$ and $\widetilde{\gamma}_s - \eta\theta_\eta$, we substitute $\psi = \mathcal{Z}_{\mathrm{re}}$ and $\psi = i\mathcal{Z}_{\mathrm{im}}$ to (6.27), respectively. In case of $\psi = \mathcal{Z}_{\mathrm{re}}$, we get

$$
\begin{aligned}
&\Big(\frac{\lambda_s}{\lambda} + b\Big)\Big((\Lambda Q^{(\eta)}, \mathcal{Z}_{\mathrm{re}})_r + O(\|\epsilon\|_{\dot H^1_m})\Big) \\
&\qquad \approx -(\epsilon, [\mathcal{L}_{Q^{(\eta)}} i\mathcal{Z}_{\mathrm{re}} + \eta\theta_\eta i\mathcal{Z}_{\mathrm{re}} + \eta^2\tfrac{i|y|^2\mathcal{Z}_{\mathrm{re}}}{4}]_b)_r - (\widetilde{\gamma}_s - \eta\theta_\eta)(\epsilon, [i\mathcal{Z}_{\mathrm{re}}]_b)_r
\end{aligned}
$$

up to error (6.34). In case of $\psi = i\mathcal{Z}_{\mathrm{im}}$, we get

$$
\begin{aligned}
&(\widetilde{\gamma}_s - \eta\theta_\eta)\big((Q^{(\eta)}, \mathcal{Z}_{\mathrm{im}})_r + O(\|\epsilon\|_{\dot H^1_m})\big) \\
&\qquad \approx -(\epsilon, [\mathcal{L}_{Q^{(\eta)}}\mathcal{Z}_{\mathrm{im}} + \eta\theta_\eta\mathcal{Z}_{\mathrm{im}} + \eta^2\tfrac{|y|^2\mathcal{Z}_{\mathrm{im}}}{4}]_b)_r - \Big(\frac{\lambda_s}{\lambda} + b\Big)(\epsilon, [i\Lambda\mathcal{Z}_{\mathrm{im}}]_b)_r
\end{aligned}
$$

up to error (6.34). Recall that $(\Lambda Q, \mathcal{Z}_{\mathrm{re}})_r = (Q, \mathcal{Z}_{\mathrm{im}})_r = 1$. If we *crudely* estimate the inner products containing ϵ by $\|\epsilon\|_{\dot H^1_m}$ and expand all the errors of (6.34), the modulation estimate (6.28) follows.

We turn to the degeneracy estimate (6.32). We substitute $\psi = Q^{(\eta)}$ into (6.27). Using the fact that $iQ^{(\eta)}$ lies in the kernel of the operator $\mathcal{L}_{Q^{(\eta)}} + \eta\theta_\eta + \eta^2\frac{|y|^2}{4}$ (see (4.24)), i.e.

$$
\mathcal{L}_{Q^{(\eta)}} iQ^{(\eta)} + \eta\theta_\eta iQ^{(\eta)} + \eta^2\tfrac{|y|^2}{4} iQ^{(\eta)} = 0,
$$

modulation estimate (6.28), and error estimate (6.34), we get

$$|\partial_s(\epsilon, Q_b^{(\eta)})_r| \lesssim \alpha^*\lambda^2(\lambda^{m+1}|\log\lambda| + \eta) + (\alpha^*\lambda + \|\epsilon\|_{\dot H^1_m})\|\epsilon\|_{\dot H^1_m}.$$

We then integrate the flow backward in time with $\epsilon(0) = 0$ and use (2.12) to get

$$\begin{aligned}|(\epsilon, Q_b^{(\eta)})_r| &\lesssim \int_t^0 \alpha^*(\lambda^{m+1}|\log\lambda| + \eta) + (\alpha^*\lambda^{-1} + 1)\|\epsilon\|_{\dot H^1_m}\,dt' \\ &\lesssim \alpha^*\lambda(\lambda^{m+1}|\log\lambda| + \eta) + (\alpha^* + \lambda)\mathcal{T}^{(\eta,\frac{5}{4})}_{\dot H^1_m}[\epsilon].\end{aligned}$$

This completes the proof of (6.32). □

6.4. L^2-bound of ϵ

The coercivity estimate (3.18) controls only $\dot H^1_m$-norm of ϵ. In order to close our bootstrap argument, we need to control L^2_m-norm of ϵ. Indeed, we will use Lyapunov/virial functional method. Since the virial correction is on $\dot H^{\frac12}$ level, we are required to estimate L^2_m-norm of ϵ.

As we will estimate $\|\epsilon(t)\|_{L^2}$ by integrating the flow backward in time (note that $\epsilon(0) = 0$), it is natural to bound by the time maximal function (recall Section 2.3). In this step, we use the Strichartz estimates.

LEMMA 6.15 (L^2-estimate of ϵ). *We have*[2]

$$\|\epsilon\|_{L^2} \lesssim \alpha^*\lambda^{m+2}|\log\lambda| + \alpha^*\lambda\eta + \lambda^{-\frac34}\mathcal{T}^{(\eta,\frac54)}_{\dot H^1_m}[\epsilon]. \tag{6.35}$$

REMARK 6.16. Here we lose $\lambda^{\frac34}$ from $\dot H^1_m$ bound. It is important not to lose λ^1 in order to guarantee the virial correction term to be relatively small compared to the energy functional. See Section 6.5.

REMARK 6.17. In the proof, we use the Strichartz estimates. If one merely tries to use the energy method on L^2, then one ends up with $\lambda^{-1}\mathcal{T}[\epsilon]$. The point of using the Strichartz estimates is that we can control L^p norm of ϵ for $p < 2$. Alternatively, one can apply the energy method on $\dot H^{-\frac12}$ with the embedding $L^{\frac43} \hookrightarrow \dot H^{-\frac12}$ and interpolate it with $\|\epsilon^\sharp\|_{\dot H^1_m} = \lambda^{-1}\|\epsilon\|_{\dot H^1_m}$.

REMARK 6.18. Losing $\lambda^{\frac34}$ from $\dot H^1_m$ bound is not optimal. In fact, one can combine Strichartz estimates and local energy estimate (see for instance [**4, 13**] and [**13**, Theorem 10.1])

$$\||x|^{-1}u\|_{L^2_{t,x}} \lesssim \|u_0\|_{L^2} + \||x|(i\partial_t + \Delta_m)u\|_{L^2_{t,x}}$$

to replace $\lambda^{\frac34}$ loss by $\lambda^{\frac12}$ loss. Nevertheless, this does not significantly improve the result of the paper. In particular, $\|\epsilon\|_{L^2} \lesssim \lambda^{-\frac12}\|\epsilon\|_{\dot H^1_m}$ is still *not enough* to say that the η-correction term $\frac{\eta\theta_\eta}{2}M[\epsilon]$ is relatively small compared to the energy functional. See Section 6.5.

REMARK 6.19. Compare (5.15) and (6.26). Deleting $[\frac{|y|^2}{4}Q^{(\eta)}]^\sharp_b$ terms is crucial for small m. By (5.18), we are able to delete $[\frac{|y|^2}{4}Q^{(\eta)}]^\sharp_b$. Otherwise, we should have estimated

$$\frac{1}{\lambda^2}\Big(\Big(\frac{\lambda_s}{\lambda} + b\Big)[\Lambda Q_b^{(\eta)}]^\sharp + (\widetilde{\gamma}_s - \eta\theta_\eta)Q_b^{(\eta)\sharp} - (b_s + b^2)[\tfrac{|y|^2}{4}Q^{(\eta)}]^\sharp_b\Big).$$

[2]Here the time maximal function is applied to $\|\epsilon(s(t'),\cdot)\|_{\dot H^1_m}$ for $t' \in [t, 0]$.

A slow decay of $\Lambda Q_b^{(\eta)}$ and $\frac{|y|^2}{4}Q_b^{(\eta)}$ is problematic for small m. For example, they do not belong to L^p for any $p \le 2$ if $m = 1$ and $\eta = 0$. This prevents us to obtain any useful L^2 bound of ϵ. Here the law (5.18) is really crucial. Without (5.18), we guess that one should truncate the profile $Q_b^{(\eta)}$ (say $\widetilde{Q}_b^{(\eta)}$) with carefully chosen radius, make an ansatz $u = \widetilde{Q}_b^{(\eta)\sharp} + z + \epsilon^\sharp$, and control the errors arising from $\widetilde{Q}_b^{(\eta)}$. This will generate more error terms and the analysis would become much more complicated.

PROOF. We use $\epsilon^\sharp$-equation (6.26) in the following form:

$$\begin{aligned}(6.36)\qquad i\partial_t\epsilon^\sharp + \Delta_m\epsilon^\sharp &= \frac{1}{\lambda^2}i\Big(\frac{\lambda_s}{\lambda}+b\Big)[\Lambda Q^{(\eta)}]_b^\sharp + \frac{1}{\lambda^2}(\widetilde{\gamma}_s - \eta\theta_\eta)Q_b^{(\eta)\sharp}\\ &\quad + \widetilde{R}_{Q_b^{(\eta)\sharp},z} + V_{Q_b^{(\eta)\sharp}-Q_b^\sharp}z\\ &\quad + (\mathcal{L}_w+\Delta_m)\epsilon^\sharp + R_{u-w}.\end{aligned}$$

By Strichartz estimates (Lemma 2.7) and $\epsilon(0)=0$, it suffices to estimate the RHS of (6.36) by the dual Strichartz norm.

We treat the first and second lines of (6.36). Note that

$$\|[\Lambda Q^{(\eta)}]_b^\sharp\|_{L_x^{\frac{4}{3}}} + \|Q_b^{(\eta)\sharp}\|_{L_x^{\frac{4}{3}}} \lesssim \lambda^{\frac12}.$$

Combining this with the modulation estimate (6.28) and the maximal estimate (2.12), we get

$$\begin{aligned}&\Big\|\frac{1}{\lambda^2}\Big(i\Big(\frac{\lambda_s}{\lambda}+b\Big)[\Lambda Q^{(\eta)}]_b^\sharp + \widetilde{\gamma}_s Q_b^{(\eta)\sharp}\Big)\Big\|_{L^{\frac43}_{[t,0],x}}\\ &\lesssim \lambda^{-\frac34}\Big(\mathcal{T}^{(\eta,\frac54)}\Big[\frac{\lambda_s}{\lambda}+b\Big] + \mathcal{T}^{(\eta,\frac54)}[\widetilde{\gamma}_s-\eta\theta_\eta]\Big)\\ &\lesssim \lambda^{-\frac34}(\alpha^*\lambda^{m+3}|\log\lambda| + \alpha^*\lambda^2\eta + \mathcal{T}^{(\eta,\frac54)}_{\dot H^1_m}[\epsilon]).\end{aligned}$$

The second line of the RHS of (6.36) is estimated as

$$\|\widetilde{R}_{Q_b^\sharp,z} + V_{Q_b^{(\eta)\sharp}-Q_b^\sharp}z\|_{L^1_{[t,0]}L^2_x} \lesssim \alpha^*\lambda^{m+2}|\log\lambda| + \alpha^*\lambda\eta.$$

To complete the proof, it only remains to show

$$\|(\mathcal{L}_w+\Delta_m)\epsilon^\sharp + R_{u-w}\|_{L^{\frac43}_{[t,0],x}} \lesssim \lambda^{-\frac34}\mathcal{T}^{(\eta,\frac54)}_{\dot H^1_m}[\epsilon].$$

We note that

$$\begin{aligned}&(\mathcal{L}_w+\Delta_m)\epsilon^\sharp + R_{u-w}\\ &= \sum_{\substack{\psi_1,\psi_2,\psi_3\in\{Q_b^{(\eta)\sharp},z,\epsilon^\sharp\}:\\ \#\{i:\psi_i=\epsilon^\sharp\}\ge1}}[\mathcal{N}_{3,0}+\mathcal{N}_{3,1}+\mathcal{N}_{3,2}] + \sum_{\substack{\psi_1,\dots,\psi_5\in\{Q_b^{(\eta)\sharp},z,\epsilon^\sharp\}:\\ \#\{i:\psi_i=\epsilon^\sharp\}\ge1}}[\mathcal{N}_{5,1}+\mathcal{N}_{5,2}].\end{aligned}$$

In the above expression, the worst terms occur when only one ψ_i is $\epsilon^\sharp$ and all the other ψ_is are $Q_b^{(\eta)\sharp}$ because the following estimate saturates when $\psi_i = Q_b^{(\eta)\sharp}$:

$$\|\psi\|_{L^p_x} \lesssim \lambda^{\frac2p-1};\qquad \forall\psi\in\{Q_b^{(\eta)\sharp},z,\epsilon^\sharp\},\ p\ge2.$$

Our main tools are $L^{\frac{4}{3}}$-estimate (Lemma 2.6) and maximal function estimate (2.12). By (2.33),

$$\Big\| \sum_{\substack{\psi_1,\psi_2,\psi_3 \in \{Q_b^{(\eta)\sharp}, z, \epsilon^\sharp\}: \\ \#\{i:\psi_i=\epsilon^\sharp\}\geq 1}} [\mathcal{N}_{3,0}+\mathcal{N}_{3,1}+\mathcal{N}_{3,2}] \Big\|_{L_x^{\frac{4}{3}}}$$

$$\lesssim \|\epsilon^\sharp\|_{\dot H_m^1} \sum_{\psi_1,\psi_2 \in \{Q_b^{(\eta)\sharp}, z, \epsilon^\sharp\}} \|\psi_1\|_{L^2}\|\psi_2\|_{L^4} \lesssim \lambda^{-\frac{3}{2}}\|\epsilon\|_{\dot H_m^1}.$$

Applying (2.12), we get

$$\Big\| \sum_{\substack{\psi_1,\psi_2,\psi_3 \in \{Q_b^{(\eta)\sharp}, z, \epsilon^\sharp\}: \\ \#\{i:\psi_i=\epsilon^\sharp\}\geq 1}} [\mathcal{N}_{3,0}+\mathcal{N}_{3,1}+\mathcal{N}_{3,2}] \Big\|_{L_{[t,0],x}^{\frac{4}{3}}} \lesssim \lambda^{-\frac{3}{4}} \mathcal{T}_{\dot H_m^1}^{(\eta,\frac{5}{4})}[\epsilon].$$

Treating $\mathcal{N}_{5,1}$ and $\mathcal{N}_{5,2}$ terms is similar; this time one uses (2.34) and scaling relation (2.35). We omit the details. This ends the proof. □

6.5. Lyapunov/virial functional

So far, we have seen that how $\dot H_m^1$ norm on ϵ controls variations of modulation parameters and also L^2 norm of ϵ. In this section, in order to close our bootstrap, we will control $\dot H_m^1$ norm of ϵ using the Lyapunov functional. Our goal is to find a Lyapunov functional $\mathcal{I}$ such that $\mathcal{I}$ is coercive and $\partial_s \mathcal{I}$ is almost positive. For instance, if we have

$$\begin{cases} \widetilde{\mathcal{I}} \sim \|\epsilon\|_{\dot H_m^1}^2, \\ \partial_s \widetilde{\mathcal{I}} \geq -o(\lambda)\|\epsilon\|_{\dot H_m^1}^2, \end{cases} \tag{6.37}$$

and assume power-type bound $\|\epsilon\|_{\dot H_m^1} \lesssim \lambda^\ell \sim |t|^\ell$ with $\ell > 0$, then by FTC

$$\|\epsilon\|_{\dot H_m^1}^2 \lesssim \widetilde{\mathcal{I}}(t) \lesssim o(1)\cdot \int_t^{t_0} \lambda^{-1}\|\epsilon\|_{\dot H_m^1}^2 dt' \lesssim o(1)\lambda^{2\ell}.$$

Note that we have used $\epsilon(t_0)=0$ and hence $\widetilde{\mathcal{I}}(t_0)=0$. As we integrate from 0 to t (backward in time), positive terms of $\partial_s\widetilde{\mathcal{I}}$ are safe.

In view of the coercivity, a natural candidate for $\widetilde{\mathcal{I}}$ would be the energy functional E. In general, for a functional $\mathcal{A}$, we define its quadratic (and higher) part of $\mathcal{A}[w^\flat + \cdot]$ by

$$\mathcal{A}_{w^\flat}^{(\mathrm{qd})}[\epsilon] := \mathcal{A}[w^\flat+\epsilon] - \mathcal{A}[w^\flat] - (\frac{\delta\mathcal{A}}{\delta u}\Big|_{u=w^\flat}, \epsilon)_r. \tag{6.38}$$

We first start with the energy functional $E_{w^\flat}^{(\mathrm{qd})}[\epsilon]$. It turns out that, however, $\partial_s E_{w^\flat}^{(\mathrm{qd})}[\epsilon]$ does not enjoy satisfactory lower bound. To resolve the difficulty, we make two modifications from $E_{w^\flat}^{(\mathrm{qd})}[\epsilon]$. Firstly, we will observe that $\partial_s E_{w^\flat}^{(\mathrm{qd})}[\epsilon]$ contains $\frac{2\lambda_s}{\lambda}E_{w^\flat}^{(\mathrm{qd})}[\epsilon]$. Note that $\frac{2\lambda_s}{\lambda}\approx -2b$ can only be controlled by λ, so $\frac{2\lambda_s}{\lambda}E_{w^\flat}^{(\mathrm{qd})}[\epsilon]$ is neither positive nor small. Thus it is natural to rescale it to hide $\frac{2\lambda_s}{\lambda}E_{w^\flat}^{(\mathrm{qd})}[\epsilon]$ using

$$\lambda^2\partial_s(\lambda^{-2}E_{w^\flat}^{(\mathrm{qd})}[\epsilon]) = \partial_s E_{w^\flat}^{(\mathrm{qd})}[\epsilon] - \frac{2\lambda_s}{\lambda}E_{w^\flat}^{(\mathrm{qd})}[\epsilon].$$

Secondly, there is another non-perturbative effect both from scalings and rotations. In view of (6.25), ϵ actually evolves under

$$i\partial_s\epsilon - \mathcal{L}_{w^b}\epsilon + ib\Lambda\epsilon - \eta\theta_\eta\epsilon \approx 0.$$

To incorporate this, we add a correction $b\Phi_A[\epsilon] + \frac{\eta\theta_\eta}{2}M[\epsilon]$ to our energy functional $E^{(\mathrm{qd})}_{w^b}[\epsilon]$. The former term $b\Phi_A[\epsilon]$ is a localized virial correction, and the latter term $\frac{\eta\theta_\eta}{2}M[\epsilon]$ is a mass correction. We will choose

$$\mathcal{I}_A := \lambda^{-2}(E^{(\mathrm{qd})}_{w^b}[\epsilon] + b\Phi_A[\epsilon] + \tfrac{\eta\theta_\eta}{2}M[\epsilon]).$$

Here, $A \geq 1$ is some large parameter to be chosen later. Still, we need a further correction on $\mathcal{I}_A$. This is due to

$$\frac{2\lambda_s}{\lambda}\cdot\frac{1}{8}\int(\Delta^2\phi_A)|\epsilon|^2$$

arising from the virial correction. As the coercivity of $\lambda^2\mathcal{I}_A$ only controls $\|\epsilon\|^2_{\dot{H}^1_m}$, this term is on the borderline of sufficient estimates. We can resolve this issue via averaging the virial correction. Eventually, we will use

$$\mathcal{I} := \frac{2}{\log A}\int_{A^{1/2}}^{A}\mathcal{I}_{A'}\frac{dA'}{A'} = \lambda^{-2}\Big(E^{(\mathrm{qd})}_{w^b}[\epsilon] + \frac{2}{\log A}\int_{A^{1/2}}^{A} b\Phi_{A'}[\epsilon]\frac{dA'}{A'} + \tfrac{\eta\theta_\eta}{2}M[\epsilon]\Big).$$

In the sequel, we denote by Err the collection of small terms satisfying

$$\text{(6.39)}\qquad |\mathrm{Err}| \lesssim \lambda\cdot\Big(A(\alpha^*\lambda^{m+2}+\alpha^*\lambda\eta)^2 + (\alpha^* + o_{A\to\infty}(1) + A\lambda^{\frac{1}{4}})(\mathcal{T}^{(\eta,\frac{5}{4})}_{\dot{H}^1_m}[\epsilon])^2\Big).$$

This error may differ line by line. Roughly speaking, Err can be thought of as $o(\lambda)\|\epsilon\|^2_{\dot{H}^1_m}$ in (6.37) under rough substitutions $\|\epsilon\|_{L^2} \lesssim \lambda^{-\frac{3}{4}}\|\epsilon\|_{\dot{H}^1_m}$ and (5.26).

LEMMA 6.20 (Estimates of $E^{(\mathrm{qd})}_{w^b}$). *We have coercivity*

$$\text{(6.40)}\qquad E^{(\mathrm{qd})}_{w^b}[\epsilon] + O\big((\alpha^*+\lambda^{\frac{1}{2}})\|\epsilon\|^2_{\dot{H}^1_m}\big) \sim \|\epsilon\|^2_{\dot{H}^1_m}.$$

We have the derivative identity

$$\text{(6.41)}\qquad \begin{aligned}\partial_s E^{(\mathrm{qd})}_{w^b}[\epsilon] &= \frac{\lambda_s}{\lambda}\{2E^{(\mathrm{qd})}_{w^b}[\epsilon] - (R_{u^b-w^b}, \Lambda Q_b^{(\eta)})_r\} \\ &\quad - \gamma_s(\mathcal{L}_{w^b}\epsilon + R_{u^b-w^b}, i\epsilon)_r \\ &\quad + \Big(\mathcal{L}_{Q_b^{(\eta)}}\epsilon, (\frac{\lambda_s}{\lambda}+b)[\Lambda Q^{(\eta)}]_b - (\tilde{\gamma}_s - \eta\theta_\eta)iQ_b^{(\eta)}\Big)_r + \mathrm{Err}.\end{aligned}$$

We can formally write (6.41) *as*[3]

$$\text{(6.42)}\qquad \begin{aligned}\partial_s E^{(\mathrm{qd})}_{w^b}[\epsilon] &= \frac{\lambda_s}{\lambda}\{(R_{u^b-w^b}, \Lambda z^b)_r + (\mathcal{L}_{w^b}\epsilon + R_{u^b-w^b}, \Lambda\epsilon)_r\} \\ &\quad - \gamma_s(\mathcal{L}_{w^b}\epsilon + R_{u^b-w^b}, i\epsilon)_r \\ &\quad + \Big(\mathcal{L}_{Q_b^{(\eta)}}\epsilon, (\frac{\lambda_s}{\lambda}+b)[\Lambda Q^{(\eta)}]_b - (\tilde{\gamma}_s - \eta\theta_\eta)iQ_b^{(\eta)}\Big)_r + \mathrm{Err}.\end{aligned}$$

[3] As we do not assume decay properties on z^b and ϵ, the terms $(R_{u^b-w^b}, \Lambda z^b)_r$ and $(\mathcal{L}_{w^b}\epsilon + R_{u^b-w^b}, \Lambda\epsilon)_r$ do not make sense by themselves.

PROOF. To show (6.40), we recall the expression (2.22). By definition (6.38), we have

$$E_{w^\flat}^{(\mathrm{qd})}[\epsilon] = \frac{1}{2}\int \Big(|\partial_r \epsilon|^2 + \frac{m^2}{r^2}|\epsilon|^2\Big)$$
$$- \sum_{\substack{\psi_1,\dots,\psi_4\in\{w^\flat,\epsilon\}:\\ \#\{i:\psi_i=\epsilon\}\geq 2}} [\frac{1}{4}\mathcal{M}_{4,0} + \frac{m}{2}\mathcal{M}_{4,1}] + \frac{1}{8}\sum_{\substack{\psi_1,\dots,\psi_6\in\{w^\flat,\epsilon\}:\\ \#\{i:\psi_i=\epsilon\}\geq 2}} \mathcal{M}_6.$$

Using the multilinear estimates (Lemma 2.2) with $\|w^\flat - Q\|_{L^2} \lesssim \alpha^* + \lambda|\log\lambda|^{\frac{1}{2}}$ (see (4.1)), we can replace $w^\flat$ by Q:

$$E_{w^\flat}^{(\mathrm{qd})}[\epsilon] = E_Q^{(\mathrm{qd})}[\epsilon] + O\big((\alpha^* + \lambda|\log\lambda|^{\frac{1}{2}})\|\epsilon\|_{\dot H_m^1}^2\big).$$

Using the expansion of energy (3.5), bootstrap hypothesis $\|\epsilon\|_{L^2} \leq \lambda^{\frac{1}{2}}$, and multilinear estimate (Lemma 2.2), we have

$$E_{w^\flat}^{(\mathrm{qd})}[\epsilon] = \frac{1}{2}\int |L_Q\epsilon|^2 + O\big((\alpha^* + \lambda^{\frac{1}{2}})\|\epsilon\|_{\dot H_m^1}^2\big).$$

We then apply the coercivity (3.18) to conclude (6.40).

We turn to compute $\partial_s E_{w^\flat}^{(\mathrm{qd})}[\epsilon]$. If we perform the computation at (s,y)-scalings, we encounter unbounded-looking terms. For example, $\partial_s w^\flat$ and $\partial_s\epsilon$ contain $\frac{\lambda_s}{\lambda}\Lambda z^\flat$ and $\frac{\lambda_s}{\lambda}\Lambda\epsilon$, respectively. These terms are not in L^2 as we do not assume decay on z and ϵ. This issue can be resolved if we perform the computation at (t,x)-scalings instead. But we first start computing $\partial_s E_{w^\flat}^{(\mathrm{qd})}[\epsilon]$ at (s,y)-scalings to observe a crucial cancellation and rearrange errors. We estimate errors in (s,y)-scalings rigorously. Later, we redo the computation of $\partial_s E_{w^\flat}^{(\mathrm{qd})}[\epsilon] = \lambda^2\partial_t(\lambda^2 E_w^{(\mathrm{qd})}[\epsilon^\sharp])$ in (t,x)-scalings to justify the formal computation.

We start with the formal computation

$$\partial_s E_{w^\flat}^{(\mathrm{qd})}[\epsilon] = \Big(\frac{\delta E}{\delta u}\Big|_{u=w^\flat+\epsilon} - \frac{\delta E}{\delta u}\Big|_{u=w^\flat} - \frac{\delta^2 E}{\delta u^2}\Big|_{u=w^\flat}\epsilon, \partial_s w^\flat\Big)_r$$
$$+ \Big(\frac{\delta E}{\delta u}\Big|_{u=w^\flat+\epsilon} - \frac{\delta E}{\delta u}\Big|_{u=w^\flat}, \partial_s\epsilon\Big)_r.$$

Note that

$$\frac{\delta E}{\delta u}\Big|_{u=w^\flat+\epsilon} - \frac{\delta E}{\delta u}\Big|_{u=w^\flat} = \mathcal{L}_{w^\flat}\epsilon + R_{u^\flat - w^\flat},$$
$$\frac{\delta^2 E}{\delta u^2}\Big|_{u=w^\flat}\epsilon = \mathcal{L}_{w^\flat}\epsilon.$$

Thus

$$\partial_s E_{w^\flat}^{(\mathrm{qd})}[\epsilon] = (R_{u^\flat - w^\flat}, \partial_s w^\flat)_r + (\mathcal{L}_{w^\flat}\epsilon + R_{u^\flat - w^\flat}, \partial_s\epsilon)_r. \tag{6.43}$$

For the first term of the RHS of (6.43), we claim

$$(R_{u^\flat - w^\flat}, \partial_s w^\flat)_r = \frac{\lambda_s}{\lambda}(R_{u^\flat - w^\flat}, \Lambda z^\flat)_r + \mathrm{Err}.$$

To see this, decompose

$$\partial_s w^\flat = b_s(\partial_b Q_b^{(\eta)}) - iL_{z^\flat}^*\mathbf{D}_+^{(z^\flat)}z^\flat - iV_{Q_b^{(\eta)}\to z^\flat}z^\flat - i\gamma_s z^\flat + \frac{\lambda_s}{\lambda}\Lambda z^\flat.$$

If we use the estimates (6.30), (6.20), (6.21), and the energy estimates of z as

$$\|z^\flat\|_{L^2} \lesssim \alpha^*,$$
$$\|L^*_{z^\flat}\mathbf{D}_+^{(z^\flat)} z^\flat\|_{L^2} + \|V_{Q_b\to z^\flat} z^\flat\|_{L^2} \lesssim \alpha^*\lambda^2,$$

we have

$$|(R_{u^\flat - w^\flat}, \partial_s w^\flat)_r| \lesssim \lambda^2(\|\epsilon\|_{L^2}\|\epsilon\|_{\dot H^1_m} + \|\epsilon\|^2_{\dot H^1_m}) + \alpha^*\lambda\|\epsilon\|^2_{\dot H^1_m}.$$

Substituting the L^2-bound (6.35) and an obvious bound $\|\epsilon\|_{\dot H^1_m} \leq \mathcal{T}^{(\eta,\frac{5}{4})}_{\dot H^1_m}[\epsilon]$, the claim follows.

We turn to the second term of the RHS of (6.43). We compute using the ϵ-equation (6.25) that

$$(\mathcal{L}_{w^\flat}\epsilon + R_{u^\flat - w^\flat}, \partial_s\epsilon)_r = (6.45) + (6.46) + (6.47) + (6.48), \tag{6.44}$$

where

$$(\mathcal{L}_{w^\flat}\epsilon + R_{u^\flat - w^\flat}, -i\mathcal{L}_{w^\flat}\epsilon - iR_{u^\flat - w^\flat})_r \tag{6.45}$$

$$(\mathcal{L}_{w^\flat}\epsilon + R_{u^\flat - w^\flat}, \Big(\frac{\lambda_s}{\lambda} + b\Big)[\Lambda Q^{(\eta)}]_b - (\widetilde{\gamma}_s - \eta\theta_\eta)iQ_b^{(\eta)})_r \tag{6.46}$$

$$(\mathcal{L}_{w^\flat}\epsilon + R_{u^\flat - w^\flat}, \frac{\lambda_s}{\lambda}\Lambda\epsilon - \gamma_s i\epsilon)_r \tag{6.47}$$

$$(\mathcal{L}_{w^\flat}\epsilon + R_{u^\flat - w^\flat}, -iV_{Q_b^{(\eta)} - Q_b} z^\flat - i\widetilde{R}_{Q_b^{(\eta)}, z^\flat})_r. \tag{6.48}$$

One crucial observation is that (6.45) *vanishes*. This is one of the reason why we work on (s,y)-scalings first.

We turn to (6.46). We claim that

$$\begin{aligned}
&\Big(\mathcal{L}_{w^\flat}\epsilon + R_{u^\flat - w^\flat}, \big(\frac{\lambda_s}{\lambda} + b\big)[\Lambda Q^{(\eta)}]_b - (\widetilde{\gamma}_s - \eta\theta_\eta)iQ_b^{(\eta)}\Big)_r \\
&= \Big(\frac{\lambda_s}{\lambda} + b\Big)(\mathcal{L}_{Q_b^{(\eta)}}\epsilon, [\Lambda Q^{(\eta)}]_b)_r + (\widetilde{\gamma}_s - \eta\theta_\eta)(\mathcal{L}_{Q_b^{(\eta)}}\epsilon, -iQ_b^{(\eta)})_r + \mathrm{Err}.
\end{aligned}$$

To show this claim, we observe for $\psi \in \{\Lambda Q^{(\eta)}, -iQ^{(\eta)}\}$ that

$$\begin{aligned}
(\mathcal{L}_{w^\flat}\epsilon + R_{u^\flat - w^\flat}, \psi_b)_r &= (\mathcal{L}_{Q_b^{(\eta)}}\epsilon, \psi_b)_r + ((\mathcal{L}_{w^\flat} - \mathcal{L}_{Q_b^{(\eta)}})\epsilon, \psi_b)_r + (R_{u^\flat - w^\flat}, \psi_b)_r \\
&= (\mathcal{L}_{Q_b^{(\eta)}}\epsilon, \psi_b)_r + O\big((\alpha^*\lambda + \lambda^{\frac{3}{2}})\|\epsilon\|_{\dot H^1_m}\big),
\end{aligned}$$

where we used $\|\psi\|_{L^2} \lesssim 1$, (6.22), (6.20), and the weak bootstrap hypothesis (5.25). Applying the modulation estimate (6.28), the claim follows. This ends the treatment of (6.46).

We will keep (6.47). This is a non-perturbative term.

Finally, (6.48) is considered as an error. We estimate using (6.19) to have

$$|(\mathcal{L}_{w^\flat}\epsilon + R_{u^\flat - w^\flat}, iV_{Q_b^{(\eta)} - Q_b} z^\flat)_r| \lesssim \|\epsilon\|_{\dot H^1_m}\|V_{Q_b^{(\eta)} - Q_b} z^\flat\|_{H^1_m} \lesssim \lambda(\alpha^*\lambda\eta)\|\epsilon\|_{\dot H^1_m}.$$

We now claim that

$$|(\mathcal{L}_{w^\flat}\epsilon + R_{u^\flat - w^\flat}, i\widetilde{R}_{Q_b^{(\eta)}, z^\flat})_r| \lesssim \lambda(\alpha^*\lambda^{m+2})\|\epsilon\|_{\dot H^1_m}. \tag{6.49}$$

Using $\|w^\flat - Q_b\|_{L^2} \lesssim \alpha^* + \lambda|\log\lambda|^{\frac{1}{2}}$, (6.7), and (6.8), it suffices to show

$$|(\mathcal{L}_{Q_b}\epsilon, i\widetilde{R}_{Q_b^{(\eta)}, z^\flat})_r| \lesssim \alpha^*\lambda^{m+3}\|\epsilon\|_{\dot H^1_m}.$$

By Lemma 6.6, it suffices to show

$$|(\mathcal{L}_{Q_b}\epsilon, [iR_2]_b)_r| \lesssim \alpha^* \lambda^{m+3} \|\epsilon\|_{\dot{H}^1_m}.$$

We now manipulate using (4.3)

$$\begin{aligned}\mathcal{L}_{Q_b}(iR_2)_b &= [L_Q^* L_Q(iR_2)]_b + ib\Lambda[iR_2]_b - b^2\tfrac{|y|^2}{4}[iR_2]_b \\ &= [L_{Q_b}^* - ib\tfrac{r}{2}][L_Q(iR_2)]_b + ib\Lambda[iR_2]_b + b^2\tfrac{|y|^2}{4}[iR_2]_b.\end{aligned}$$

We estimate using Lemma 6.6 as

$$(\epsilon, L_{Q_b}^*[L_Q(iR_2)]_b)_r \lesssim \alpha^*\lambda^{m+3}\|L_{Q_b}\epsilon\|_{L^2} \lesssim \alpha^*\lambda^{m+3}\|\epsilon\|_{\dot{H}^1_m}.$$

All the other terms are treated crudely. We use

$$\partial_r(e^{-ib\frac{r^2}{4}}) = -ib\tfrac{r}{2}e^{-ib\frac{r^2}{4}}$$

to integrate by parts

$$\begin{aligned}&(\epsilon, -ib\tfrac{r}{2}[L_Q(iR_2)]_b)_r \\ &= ([-\partial_r - \tfrac{1}{r}]\epsilon, [L_Q(iR_2)]_b)_r + (\epsilon, [\partial_r L_Q(iR_2)]_b)_r \lesssim \alpha^*\lambda^{m+3}\|\epsilon\|_{\dot{H}^1_m}.\end{aligned}$$

We then estimate

$$(\epsilon, ib\Lambda[iR_2]_b)_r = (-\Lambda\epsilon, ib[iR_2]_b)_r \lesssim \|\epsilon\|_{\dot{H}^1_m}\|brR_2\|_{L^2}.$$

We then integrate by parts

$$\begin{aligned}(\epsilon, b^2\tfrac{r^2}{4}[iR_2]_b)_r &= ((-\partial_r - \tfrac{2}{r})\epsilon, ib\tfrac{r}{2}[iR_2]_b)_r + (\epsilon, ib\tfrac{r}{2}[\partial_r(iR_2)]_b)_r \\ &\lesssim \|\epsilon\|_{\dot{H}^1_m}(\|brR_2\|_{L^2} + \|br^2\partial_r R_2\|_{L^2}).\end{aligned}$$

This completes the proof of (6.49). This ends the formal derivation of (6.42).

Now we make all of the above computations rigorous. The terms $(R_{u^\flat - w^\flat}, \Lambda z^\flat)_r$ and $(\mathcal{L}_{w^\flat}\epsilon + R_{u^\flat - w^\flat}, \Lambda\epsilon)_r$ do not make sense by themselves. However, the sum of two terms $(R_{u^\flat - w^\flat}, \Lambda z^\flat)_r$ and $(\mathcal{L}_{w^\flat}\epsilon + R_{u^\flat - w^\flat}, \Lambda\epsilon)_r$ does make sense from the following formal computation:

$$\begin{aligned}&(R_{u^\flat - w^\flat}, \Lambda z^\flat)_r + (\mathcal{L}_{w^\flat}\epsilon + R_{u^\flat - w^\flat}, \Lambda\epsilon)_r \\ &= \{(R_{u^\flat - w^\flat}, \Lambda w^\flat)_r + (\mathcal{L}_{w^\flat}\epsilon + R_{u^\flat - w^\flat}, \Lambda\epsilon)_r\} - (R_{u^\flat - w^\flat}, \Lambda Q_b^{(\eta)})_r \\ &= \partial_\lambda|_{\lambda=1} E^{(\mathrm{qd})}_{(w^\flat)_\lambda}[\epsilon_\lambda] - (R_{u^\flat - w^\flat}, \Lambda Q_b^{(\eta)})_r \\ &= 2E^{(\mathrm{qd})}_{w^\flat}[\epsilon] - (R_{u^\flat - w^\flat}, \Lambda Q_b^{(\eta)})_r,\end{aligned}$$

where we temporarily denoted $f_\lambda(y) := \lambda f(\lambda y)$ for $f \in \{w^\flat, \epsilon\}$. This is how we obtain (6.41) from the formal one (6.42).

If one wants to avoid unbounded-looking quantities, we can compute in (t, x)-scalings

$$\partial_s E^{(\mathrm{qd})}_{w^\flat}[\epsilon] = \lambda^2\partial_t(\lambda^2 E^{(\mathrm{qd})}_w[\epsilon^\sharp])$$

instead. Here, we use equations

$$\begin{aligned}i\partial_t w &= i\frac{1}{\lambda^2}[b_s\partial_b Q_b^{(\eta)} - \frac{\lambda_s}{\lambda}\Lambda Q_b^{(\eta)} + i\gamma_s Q_b^{(\eta)}]^\sharp + L_z^*\mathbf{D}_+^{(z)}z + V_{Q_b^\sharp \to z}z, \\ i\partial_t\epsilon^\sharp &= \mathcal{L}_w\epsilon^\sharp + \frac{1}{\lambda^2}i\Big(\frac{\lambda_s}{\lambda} + b\Big)[\Lambda Q^{(\eta)}]_b^\sharp + \frac{1}{\lambda^2}(\widetilde{\gamma}_s - \eta\theta_\eta)Q_b^{(\eta)\sharp} \\ &\quad + \widetilde{R}_{Q_b^\sharp, z} + V_{Q_b^{(\eta)\sharp} - Q_b^\sharp}z + R_{u-w},\end{aligned}$$

where we used dynamic rescaling for $i\partial_t Q_b^{(\eta)\sharp}$, (zCSS) for $i\partial_t z$, and (6.26) for $i\partial_t \epsilon^\sharp$. We then obtain (6.41) by lowering to the (s,y)-scale and estimating errors as before. □

The RHS of (6.41) contains non-perturbative terms. Indeed, we roughly have

$$\frac{2\lambda_s}{\lambda} E_{w^\flat}^{(\mathrm{qd})}[\epsilon] \approx -2bE_{w^\flat}^{(\mathrm{qd})}[\epsilon] \sim -b\|\epsilon\|_{\dot H_m^1}^2,$$

$$\Big|\frac{\lambda_s}{\lambda}(R_{u^\flat - w^\flat}, \Lambda Q_b^{(\eta)})_r\Big| \lesssim b\|R_{u^\flat - w^\flat}\|_{L^2} \lesssim b\|\epsilon\|_{\dot H_m^1}^2,$$

These are at the borderline of the acceptable error size Err. Next, from $\gamma_s \approx \eta\theta_\eta$,

$$|\gamma_s(\mathcal{L}_{w^\flat}\epsilon + R_{u^\flat - w^\flat}, i\epsilon)_r| \lesssim \eta\|\epsilon\|_{\dot H_m^1}^2,$$

is also at the borderline. The remaining terms are also at the borderline. We know from the modulation estimates that $|\frac{\lambda_s}{\lambda} + b| + |\widetilde{\gamma}_s - \eta\theta_\eta| \lesssim \|\epsilon\|_{\dot H_m^1}$. The problem is that $(\mathcal{L}_{Q_b^{(\eta)}}\epsilon, \psi_b)_r$ for $\psi \in \{\Lambda Q^{(\eta)}, iQ^{(\eta)}\}$ cannot be estimated better than $\lambda\|\epsilon\|_{\dot H_m^1}$; from (4.3), (4.24), we have

$$(\epsilon, \mathcal{L}_{Q_b^{(\eta)}} iQ_b^{(\eta)})_r = (\epsilon, -\eta\theta_\eta iQ_b^{(\eta)} - b\Lambda Q_b^{(\eta)})_r + (\epsilon, -(b^2+\eta^2)\tfrac{|y|^2}{4} iQ_b^{(\eta)})_r, \tag{6.50}$$

where the term $(\epsilon, -\eta\theta_\eta iQ_b^{(\eta)} - b\Lambda Q_b^{(\eta)})_r$ is at best estimated by $\lambda\|\epsilon\|_{\dot H_m^1}$.[4]

We can resolve this issue by modifying the functional $E_{w^\flat}^{(\mathrm{qd})}[\epsilon]$. Firstly, we hide

$$\frac{2\lambda_s}{\lambda} E_{w^\flat}^{(\mathrm{qd})}[\epsilon]$$

by rescaling the functional as

$$\lambda^2\partial_s(\lambda^{-2}E_{w^\flat}^{(\mathrm{qd})}[\epsilon]) = -\frac{2\lambda_s}{\lambda}E_{w^\flat}^{(\mathrm{qd})}[\epsilon] + \partial_s E_{w^\flat}^{(\mathrm{qd})}[\epsilon].$$

For the remaining three terms, we can weaken them by adding virial and mass corrections. In view of (6.25), ϵ actually evolves under

$$i\partial_s\epsilon - \mathcal{L}_{w^\flat}\epsilon + ib\Lambda\epsilon - \eta\theta_\eta\epsilon \approx 0.$$

The energy functional $E_{w^\flat}^{(\mathrm{qd})}[\epsilon]$ is adapted to the evolution $(i\partial_s - \mathcal{L}_{w^\flat})\epsilon \approx 0$. To take the $ib\Lambda\epsilon$ part into account, it is natural to introduce the virial term

$$b\Phi[\epsilon] = \frac{b}{2}\mathrm{Im}\int \overline{\epsilon}\cdot r\partial_r\epsilon$$

so that $\frac{\delta\Phi}{\delta u} = -i\Lambda u$. To take $\eta\theta_\eta\epsilon$ into account, we introduce the mass term

$$\tfrac{\eta\theta_\eta}{2}M[\epsilon] = \tfrac{\eta\theta_\eta}{2}\int |\epsilon|^2.$$

[4] In fact, one may cleverly choose orthogonality conditions to overcome this issue. For sake of simiplicity, assume $\eta = 0$ and m is large so that Q has rapid decay. If one chooses $\mathcal{Z}_{\mathrm{re}} = \Lambda Q$, then the first term of (6.50) vanishes up to error $b^2\|\epsilon\|_{\dot H_m^1}$, which is acceptable. Alternatively, one can choose $(\epsilon, |y|^2 Q_b)_r = (\epsilon, i\rho_b)_r = 0$ to have an improved modulation estimates; see Remark 6.13. Nonetheless, we stick to generic orthogonality conditions $(\epsilon, [\mathcal{Z}_{\mathrm{re}}]_b)_r = (\epsilon, [i\mathcal{Z}_{\mathrm{im}}]_b)_r = 0$ to emphasize the algebra (6.60) below.

We then have

$$\begin{aligned}&\partial_s(E^{(\mathrm{qd})}_{w^\flat}[\epsilon]+b\Phi[\epsilon]+\tfrac{\eta\theta_\eta}{2}M[\epsilon])\\&\approx(R_{u^\flat-w^\flat},\partial_s w^\flat)_r+(\mathcal{L}_{w^\flat}\epsilon+R_{u^\flat-w^\flat}-ib\Lambda\epsilon+\eta\theta_\eta\epsilon,\partial_s\epsilon)_r.\end{aligned}$$

Thus we observe that (6.45) is changed into

$$(\mathcal{L}_{w^\flat}\epsilon+R_{u^\flat-w^\flat}-ib\Lambda\epsilon+\eta\theta_\eta\epsilon,-i(\mathcal{L}_{w^\flat}\epsilon+R_{u^\flat-w^\flat}-ib\Lambda\epsilon+\eta\theta_\eta\epsilon))_r=0$$

and results in further cancellation in (6.46) and (6.47). More precisely, (6.46) is changed into

$$\Big(\mathcal{L}_{w^\flat}\epsilon+R_{u^\flat-w^\flat}-ib\Lambda\epsilon+\eta\theta_\eta\epsilon,\Big(\frac{\lambda_s}{\lambda}+b\Big)[\Lambda Q^{(\eta)}]_b-(\widetilde{\gamma}_s-\eta\theta_\eta)iQ_b^{(\eta)}\Big)_r,$$

and (6.47) is changed into

$$\Big(\mathcal{L}_{w^\flat}\epsilon+R_{u^\flat-w^\flat}-ib\Lambda\epsilon+\eta\theta_\eta\epsilon,(\frac{\lambda_s}{\lambda}+b)\Lambda\epsilon-(\gamma_s-\eta\theta_\eta)i\epsilon\Big)_r.$$

In view of the modulation estimates, (4.24), and (4.25), the above terms fall into the error.

However, the problem we still have is that $\Phi[\epsilon]$ and $(R_{u^\flat-w^\flat},\Lambda z^\flat)_r$ do not make sense as we do not assume decay on ϵ and $z^\flat$. It turns out that we can resolve these difficulties by making a suitable truncation of Φ. For $A\geq 1$, recall the radial weight ϕ_A in (2.20) and define

$$\Phi_A[\epsilon]:=\frac{1}{2}\int\phi_A'\mathrm{Im}(\overline{\epsilon}\partial_r\epsilon).$$

Note that Φ_A agrees with the usual virial functional if $y\leq A$ and approximates the Morawetz functional if $y\geq 2A$. We write Λ_A as

$$\frac{\delta\Phi_A}{\delta u}=-i\Lambda_A u:=-i[\phi_A'\partial_r+\tfrac{1}{2}(\Delta\phi_A)]u.$$

We formally have $\Lambda_\infty=\Lambda$. We will frequently use the estimate

$$|\Lambda_A u|\leq\min\{A,r\}(|\partial_r u|+|r^{-1}u|).$$

We will view the virial correction $b\Phi_A[\epsilon]$ as an error. The mass correction $\frac{\eta\theta_\eta}{2}M[\epsilon]$ is not perturbative, but it is positive when $\eta\geq 0$. This is one of the places where the sign condition $\eta\geq 0$ is necessary.

LEMMA 6.21 (Estimates of the correction). *We have*

$$|b\Phi_A[\epsilon]|\lesssim A\lambda\|\epsilon\|_{L^2}\|\epsilon\|_{\dot H^1_m}\tag{6.51}$$

and

$$\begin{aligned}&\partial_s(b\Phi_A[\epsilon]+\tfrac{\eta\theta_\eta}{2}M[\epsilon])\\&=(\mathcal{L}_{w^\flat}\epsilon+R_{u^\flat-w^\flat},-\frac{\lambda_s}{\lambda}\Lambda_A\epsilon+i\eta\theta_\eta\epsilon)_r\\&\quad+(-ib\Lambda_A\epsilon+\eta\theta_\eta\epsilon,\Big(\frac{\lambda_s}{\lambda}+b\Big)[\Lambda Q^{(\eta)}]_b-(\widetilde{\gamma}_s-\eta\theta_\eta)iQ_b^{(\eta)})_r)_r+\mathrm{Err}.\end{aligned}\tag{6.52}$$

PROOF. The estimate (6.51) is an easy consequence of $b\lesssim\lambda$ and the Cauchy-Schwarz inequality.

We turn to the estimate (6.52). Using $|b_s|\lesssim\lambda^2$ of (6.30), we see that

$$|b_s\Phi_A[\epsilon]|\lesssim\lambda\cdot O(A\lambda\|\epsilon\|_{L^2}\|\epsilon\|_{\dot H^1_m})$$

can be absorbed into Err. Thus

$$\partial_s(b\Phi_A[\epsilon] + \tfrac{\eta\theta_\eta}{2} M[\epsilon]) = (b\Lambda_A\epsilon + i\eta\theta_\eta\epsilon, i\partial_s\epsilon)_r + \mathrm{Err}.$$

We compute using the ϵ-equation (6.25):

$$(b\Lambda_A\epsilon + i\eta\theta_\eta\epsilon, i\partial_s\epsilon)_r = (6.53) + (6.54) + (6.55) + (6.56),$$

where

$$(b\Lambda_A\epsilon + i\eta\theta_\eta\epsilon, \mathcal{L}_{w^\flat}\epsilon + R_{u^\flat - w^\flat})_r, \tag{6.53}$$

$$(b\Lambda_A\epsilon + i\eta\theta_\eta\epsilon, i\Big(\frac{\lambda_s}{\lambda} + b\Big)[\Lambda Q^{(\eta)}]_b + (\widetilde{\gamma}_s - \eta\theta_\eta)Q_b^{(\eta)})_r, \tag{6.54}$$

$$(b\Lambda_A\epsilon + i\eta\theta_\eta\epsilon, i\frac{\lambda_s}{\lambda}\Lambda\epsilon + \gamma_s\epsilon)_r, \tag{6.55}$$

$$(b\Lambda_A\epsilon + i\eta\theta_\eta\epsilon, V_{Q_b^{(\eta)} - Q_b} z^\flat + \widetilde{R}_{Q_b^{(\eta)}, z^\flat})_r. \tag{6.56}$$

We will treat (6.53)-(6.56) term by term. Firstly, we show that (6.53) is equal to the first term of the RHS of (6.52) up to the error Err. Indeed, the difference is given by

$$\begin{aligned}
&\Big(\frac{\lambda_s}{\lambda} + b\Big)(\Lambda_A\epsilon, \mathcal{L}_{w^\flat}\epsilon + R_{u^\flat - w^\flat})_r \\
&= \Big(\frac{\lambda_s}{\lambda} + b\Big) \cdot \Big((\Lambda_A\epsilon, -\Delta_m\epsilon)_r + (\Lambda_A\epsilon, (\mathcal{L}_{w^\flat} + \Delta_m)\epsilon + R_{u^\flat - w^\flat})_r\Big) \\
&= \Big|\frac{\lambda_s}{\lambda} + b\Big| \cdot O(A\|\epsilon\|_{\dot{H}^1_m}^2),
\end{aligned}$$

where we used

$$(\Lambda_A\epsilon, -\Delta_m\epsilon)_r = \int \Big(\phi_A''|\partial_r\epsilon|^2 - \frac{\Delta^2\phi_A}{4}|\epsilon|^2 + \frac{\phi_A'}{r}\cdot\frac{m^2}{r^2}|\epsilon|^2\Big)$$

and the multilinear estimates (Lemma 2.2) with $\|\Lambda_A\epsilon\|_{L^2} \lesssim A\|\epsilon\|_{\dot{H}^1_m}$. Thus the above difference can be absorbed into Err by the modulation estimate.

We will keep the term (6.54) as it is not perturbative.

We now show that (6.55) can be absorbed into Err. We first observe by anti-symmetricity of i, Λ_A, and Λ that

$$(b\Lambda_A\epsilon + i\eta\theta_\eta\epsilon, i\frac{\lambda_s}{\lambda}\Lambda\epsilon + \gamma_s\epsilon)_r = (b\Lambda_A\epsilon, i\frac{\lambda_s}{\lambda}\Lambda\epsilon)_r.$$

We then observe that

$$\Big|(b\Lambda_A\epsilon, i\frac{\lambda_s}{\lambda}\Lambda\epsilon)_r\Big| \lesssim \lambda^2|(i[\Lambda, \Lambda_A]\epsilon, \epsilon)_r| \lesssim A\lambda^2\|\epsilon\|_{L^2}\|\epsilon\|_{\dot{H}^1_m},$$

where we used (6.30) and

$$\begin{aligned}
[\Lambda_A, \Lambda] &= -r\phi_A''\partial_r - \tfrac{1}{2}r(\Delta\phi_A)' + \phi_A'\partial_r, \\
|(i[\Lambda_A, \Lambda]\epsilon, \epsilon)_r| &= |(i(-r\phi_A'' + \phi_A')\partial_r\epsilon, \epsilon)_r| \lesssim A\|\epsilon\|_{L^2}\|\epsilon\|_{\dot{H}^1_m}.
\end{aligned}$$

Applying the L^2-bound (6.35), we conclude that (6.55) is absorbed into Err.

Finally, we show that (6.56) can be absorbed into Err:

$$\begin{aligned}
|(6.56)| &\lesssim \|b\Lambda_A\epsilon + i\eta\theta_\eta\epsilon\|_{L^2}\|V_{Q_b^{(\eta)} - Q_b} z^\flat + \widetilde{R}_{Q_b^{(\eta)}, z^\flat}\|_{L^2} \\
&\lesssim (A\lambda\|\epsilon\|_{\dot{H}^1_m} + \eta\|\epsilon\|_{L^2})(\alpha^*\lambda^{m+3}|\log\lambda| + \alpha^*\lambda^2\eta).
\end{aligned}$$

Substituting the L^2-bound (6.35) and $\eta \lesssim \lambda$, (6.56) is absorbed into Err. This completes the proof. □

Incorporating the above discussions, we define the *unaveraged Lyapunov/virial* functional by

$$\mathcal{I}_A = \lambda^{-2}(E^{(\mathrm{qd})}_{w^\flat}[\epsilon] + b\Phi_A[\epsilon] + \tfrac{\eta\theta_\eta}{2}M[\epsilon]).$$

LEMMA 6.22 (Unaveraged Lyapunov/virial functional). *We have*

$$\lambda^2\mathcal{I}_A - \tfrac{\eta\theta_\eta}{2}M[\epsilon] + O\big((\alpha^* + \lambda^{\frac{1}{2}})\|\epsilon\|^2_{\dot{H}^1_m} + A\lambda\|\epsilon\|_{L^2}\|\epsilon\|_{\dot{H}^1_m}\big) \sim \|\epsilon\|^2_{\dot{H}^1_m}, \tag{6.57}$$

$$\lambda^2\partial_s\mathcal{I}_A = 2b\Big(\mathcal{J}_A + \tfrac{\eta\theta_\eta}{2}M[\epsilon] - \frac{1}{8}\int(\Delta^2\phi_A)|\epsilon|^2\Big) + \mathrm{Err}, \tag{6.58}$$

where

$$\begin{aligned}\mathcal{J}_A := \frac{1}{2}\int \phi_A''|\partial_r\epsilon|^2 + \frac{1}{2}\int\frac{\phi_A'}{r}\cdot\frac{m^2}{r^2}|\epsilon|^2 \\ - \sum_{\substack{\psi_1,\dots,\psi_4\in\{Q,\epsilon\}:\\ \#\{i:\psi_i=\epsilon\}=2}}[\frac{1}{4}\mathcal{M}^{(A)}_{4,0} + \frac{m}{2}\mathcal{M}^{(A)}_{4,1}] + \frac{1}{8}\sum_{\substack{\psi_1,\dots,\psi_6\in\{Q,\epsilon\}:\\ \#\{i:\psi_i=\epsilon\}=2}}\mathcal{M}^{(A)}_6.\end{aligned} \tag{6.59}$$

To prove Lemma 6.22, we set aside some computations into a lemma.

LEMMA 6.23. *We have*

$$\begin{aligned}&(\mathcal{L}_{Q_b^{(\eta)}}\epsilon - ib\Lambda_A\epsilon + \eta\theta_\eta\epsilon, [\Lambda Q^{(\eta)}]_b)_r \\ &= -2\eta\theta_\eta(\epsilon, Q_b^{(\eta)})_r + o_{A\to\infty}(\lambda\|\epsilon\|_{\dot{H}^1_m}) + O(\lambda^{\frac{3}{2}}\|\epsilon\|_{\dot{H}^1_m})\end{aligned} \tag{6.60}$$

and

$$(\mathcal{L}_{Q_b^{(\eta)}}\epsilon - ib\Lambda_A\epsilon + \eta\theta_\eta\epsilon, [iQ^{(\eta)}]_b)_r = o_{A\to\infty}(\lambda\|\epsilon\|_{\dot{H}^1_m}) + O(\lambda^{\frac{3}{2}}\|\epsilon\|_{\dot{H}^1_m}). \tag{6.61}$$

PROOF. We will only prove (6.60). The proof of (6.61) is similar (it is in fact easier), one uses (4.24) instead of (4.25).

To prove (6.60), we crucially use the explicit computation (4.25). Applying the pseudoconformal phase, we obtain

$$\begin{aligned}&(\mathcal{L}_{Q_b^{(\eta)}} - ib\Lambda_A + \eta\theta_\eta)[\Lambda Q^{(\eta)}]_b \\ &= [(\mathcal{L}_{Q^{(\eta)}} + \eta\theta_\eta + \eta^2\tfrac{|y|^2}{4})(\Lambda Q^{(\eta)})]_b + ib(\Lambda - \Lambda_A)[\Lambda Q^{(\eta)}]_b - (b^2+\eta^2)\tfrac{|y|^2}{4}[\Lambda Q^{(\eta)}]_b \\ &= -2\eta\theta_\eta Q_b^{(\eta)} - 4\eta^2\tfrac{|y|^2Q_b^{(\eta)}}{4} + ib(\Lambda - \Lambda_A)[\Lambda Q^{(\eta)}]_b - (b^2+\eta^2)\tfrac{|y|^2}{4}[\Lambda Q^{(\eta)}]_b.\end{aligned}$$

Thus

$$\begin{aligned}&(\mathcal{L}_{Q_b^{(\eta)}}\epsilon - ib\Lambda_A\epsilon + \eta\theta_\eta\epsilon, [\Lambda Q^{(\eta)}]_b)_r \\ &= (\epsilon, -2\eta\theta_\eta Q_b^{(\eta)} - 4\eta^2\tfrac{|y|^2Q_b^{(\eta)}}{4} + ib(\Lambda - \Lambda_A)[\Lambda Q^{(\eta)}]_b - (b^2+\eta^2)\tfrac{|y|^2}{4}[\Lambda Q^{(\eta)}]_b)_r.\end{aligned}$$

If m is large, we can use the decay $|Q^{(\eta)}| + |\Lambda Q^{(\eta)}| \lesssim (1+r)^{-(m+2)}$ to conclude the proof.

To treat the worst case $m = 1$, however, we have to deal with the slow decay $r^2\Lambda Q^{(\eta)}$ especially when $\eta = 0$.[5] By (4.14), we have

$$\eta^2(\|r\cdot r^2Q^{(\eta)}\|_{L^2} + \|r\cdot r^2\Lambda Q^{(\eta)}\|_{L^2}) \lesssim \eta^{\frac{3}{2}}.$$

[5]For each $\eta > 0$, $r^2\Lambda Q^{(\eta)}$ in fact shows exponential decay $e^{-\eta\frac{r^2}{4}}$. However, we search for uniform estimates in η.

Thus

$$\begin{aligned}&(\mathcal{L}_{Q_b^{(\eta)}}\epsilon - ib\Lambda_A\epsilon + \eta\theta^{(\eta)}\epsilon, [\Lambda Q^{(\eta)}]_b)_r \\ &= -2\eta\theta^{(\eta)}(\epsilon, Q_b^{(\eta)})_r + (\epsilon, ib(\Lambda - \Lambda_A)[\Lambda Q^{(\eta)}]_b - b^2\tfrac{|y|^2}{4}[\Lambda Q^{(\eta)}]_b)_r + O(\eta^{\frac{3}{2}}\|\epsilon\|_{\dot H^1_m}).\end{aligned}$$

We then use the algebra

$$ib(\Lambda - \Lambda_A)[\Lambda Q^{(\eta)}]_b = ib[(\Lambda - \Lambda_A)(\Lambda Q^{(\eta)})]_b + b^2(\tfrac{r^2 - r\phi_A'(r)}{2})[\Lambda Q^{(\eta)}]_b$$

and the estimate

$$\|r\cdot(\Lambda - \Lambda_A)(\Lambda Q^{(\eta)})\|_{L^2} \lesssim \|\mathbf{1}_{r\geq A} rQ\|_{L^2} \lesssim A^{-1}$$

to further estimate

$$\begin{aligned}&(\mathcal{L}_{Q_b^{(\eta)}}\epsilon - ib\Lambda_A\epsilon + \eta\theta^{(\eta)}\epsilon, [\Lambda Q^{(\eta)}]_b)_r \\ &= -2\eta\theta^{(\eta)}(\epsilon, Q_b^{(\eta)})_r + b^2(\epsilon, \tfrac{(r^2 - 2r\phi_A'(r))}{4}[\Lambda Q^{(\eta)}]_b)_r \\ &\quad + o_{A\to\infty}(\lambda\|\epsilon\|_{\dot H^1_m}) + O(\eta^{\frac{3}{2}}\|\epsilon\|_{\dot H^1_m}).\end{aligned}$$

It now suffices to estimate the inner product $b^2(\epsilon, \frac{(r^2-2r\phi_A'(r))}{4}[\Lambda Q^{(\eta)}]_b)_r$. We decompose it into the parts $\mathbf{1}_{r\leq b^{-1/2}}$ and $\mathbf{1}_{r\geq b^{-1/2}}$ and estimate the latter using integration by parts. On one hand, we have

$$\begin{aligned}&b^2|(\epsilon, \mathbf{1}_{r\leq b^{-1/2}}\tfrac{(r^2-2r\phi_A'(r))}{4}[\Lambda Q^{(\eta)}]_b)_r| \\ &\lesssim b^2\|r^{-1}\epsilon\|_{L^2}\|\mathbf{1}_{r\leq b^{-1/2}}r^3Q\|_{L^2} \lesssim b^{\frac{3}{2}}\|\epsilon\|_{\dot H^1_m}.\end{aligned}$$

On the other hand, we integrate by parts with $\partial_r e^{-ib\frac{r^2}{4}} = -ib\frac{r}{2}\cdot e^{-ib\frac{r^2}{4}}$ to estimate

$$\begin{aligned}&b^2|(\epsilon, \mathbf{1}_{r\geq b^{-1/2}}\tfrac{(r^2-2r\phi_A'(r))}{4}[\Lambda Q^{(\eta)}]_b)_r| \\ &\lesssim b\|\epsilon\|_{\dot H^1_m}\|\mathbf{1}_{r\geq b^{-1/2}}rQ\|_{L^2} + b\|\epsilon\|_{L^\infty}([r^2Q]\big|_{r=b^{-1/2}}) \\ &\lesssim b^{\frac{3}{2}}\|\epsilon\|_{\dot H^1_m}.\end{aligned}$$

This completes the proof. □

PROOF OF LEMMA 6.22. The coercivity (6.57) follows from (6.40), (6.51). We turn to show the derivative estimate (6.58). Note that

$$\lambda^2\partial_s\mathcal{I}_A = (-\frac{2\lambda_s}{\lambda} + \partial_s)(E^{(\mathrm{qd})}_{w^\flat}[\epsilon] + b\Phi_A[\epsilon] + \tfrac{\eta\theta_\eta}{2}M[\epsilon]).$$

On one hand, by the modulation estimate, we have

$$-\frac{2\lambda_s}{\lambda}(E^{(\mathrm{qd})}_{w^\flat}[\epsilon] + b\Phi_A[\epsilon] + \tfrac{\eta\theta_\eta}{2}M[\epsilon]) = -\frac{2\lambda_s}{\lambda}E^{(\mathrm{qd})}_{w^\flat}[\epsilon] + b\eta\theta_\eta M[\epsilon] + \mathrm{Err}.$$

Unlike the term $-\frac{2b\lambda_s}{\lambda}\Phi_A[\epsilon]$, we *cannot* view $b\eta\theta_\eta M[\epsilon]$ as an error term. Thus we are crucially using the sign condition $b, \eta \geq 0$.

On the other hand, we use (6.42) and (6.52) to compute

$$\begin{aligned}&\partial_s[E^{(\mathrm{qd})}_{w^\flat}[\epsilon] + b\Phi_A + \tfrac{\eta\theta_\eta}{2}M] \\ &= \frac{\lambda_s}{\lambda}\{(R_{u^\flat - w^\flat}, \Lambda z^\flat)_r + (\mathcal{L}_{w^\flat}\epsilon + R_{u^\flat - w^\flat}, (\Lambda - \Lambda_A)\epsilon)_r\} \\ &\quad + (6.62) + (6.63) + \mathrm{Err},\end{aligned}$$

where

$$-(\gamma_s-\eta\theta_\eta)(\mathcal{L}_{w^\flat}\epsilon+R_{u^\flat-w^\flat},i\epsilon)_r, \tag{6.62}$$

$$(\mathcal{L}_{Q_b^{(\eta)}}\epsilon-ib\Lambda_A\epsilon+\eta\theta_\eta\epsilon,\Big(\frac{\lambda_s}{\lambda}+b\Big)[\Lambda Q^{(\eta)}]_b-(\widetilde{\gamma}_s-\eta\theta_\eta)iQ_b^{(\eta)})_r. \tag{6.63}$$

We show that (6.62) and (6.63) can be absorbed into Err. Indeed, (6.62) is absorbed into Err by observing

$$\begin{aligned}|\gamma_s-\eta\theta_\eta|\lesssim|\theta_{z^\flat\to Q_b^{(\eta)}}|+|\widetilde{\gamma}_s-\eta\theta_\eta|\lesssim\alpha^*\lambda^2+\|\epsilon\|_{\dot H^1_m},\\ |(\mathcal{L}_{w^\flat}\epsilon+R_{u^\flat-w^\flat},i\epsilon)_r|\lesssim\|\epsilon\|_{\dot H^1_m}^2.\end{aligned}$$

For (6.63), we use Lemma 6.23, the modulation estimate (6.28), and *degeneracy estimate* (6.32) to get

$$\begin{aligned}|(6.63)|\lesssim(\alpha^*\lambda^{m+3}|\log\lambda|+\alpha^*\lambda^2\eta+\|\epsilon\|_{\dot H^1_m})\cdot\Big(\lambda(o_{A\to\infty}(1)+\lambda^{\frac12})\|\epsilon\|_{\dot H^1_m}\\ +\eta(\alpha^*\lambda^{m+2}|\log\lambda|+\alpha^*\lambda\eta+(\alpha^*+\lambda)\mathcal{T}_{\dot H^1_m}^{(\frac54)}[\epsilon])\Big).\end{aligned}$$

Applying the elementary bound $\eta\lesssim\lambda$, we see that (6.63) is absorbed into Err.

Therefore, we are led to

$$\begin{aligned}&\lambda^2\partial_s\mathcal{I}_A-b\eta\theta_\eta M[\epsilon]\\ &=-\frac{\lambda_s}{\lambda}\Big(2E_{w^\flat}^{(\mathrm{qd})}[\epsilon]-(\mathcal{L}_{w^\flat}\epsilon+R_{u^\flat-w^\flat},(\Lambda-\Lambda_A)\epsilon)_r-(R_{u^\flat-w^\flat},\Lambda z^\flat)_r\Big)+\mathrm{Err}.\end{aligned}$$

One then integrate by parts the $(\Lambda-\Lambda_A)$ portion to get

(6.64)

$$\begin{aligned}&\lambda^2\partial_s\mathcal{I}_A-b\eta\theta_\eta M[\epsilon]\\ &=-\frac{\lambda_s}{\lambda}\Big(2E_{w^\flat}^{(A),(\mathrm{qd})}[\epsilon]-(R_{u^\flat-w^\flat},\Lambda_A z^\flat)_r+(R_{u^\flat-w^\flat},(\Lambda-\Lambda_A)Q_b^{(\eta)})_r\Big)+\mathrm{Err},\end{aligned}$$

where $E_{w^\flat}^{(A),(\mathrm{qd})}[\epsilon]$ is a deformation of $E_{w^\flat}^{(\mathrm{qd})}[\epsilon]$, which coincides $E_{w^\flat}^{(\mathrm{qd})}[\epsilon]$ when $A=\infty$ formally. The precise formula of $E_{w^\flat}^{(A),(\mathrm{qd})}[\epsilon]$ is given by the quadratic and higher parts of $E^{(A)}[w^\flat+\epsilon]$, where

$$\begin{aligned}E^{(A)}[u^\flat]=\frac12\int\phi_A''|\partial_r u^\flat|^2-\frac18\int(\Delta^2\phi_A)|u^\flat|^2\\ -\frac14\int\frac{\Delta\phi_A}{2}|u^\flat|^4+\frac12\int\frac{\phi_A'}{r}\Big(\frac{m+A_\theta[u^\flat]}{r}\Big)^2|u^\flat|^2.\end{aligned}$$

More precisely,

(6.65)

$$\begin{aligned}E_{w^\flat}^{(A),(\mathrm{qd})}[\epsilon]=\frac12\int\phi_A''|\partial_r\epsilon|^2-\frac18\int(\Delta^2\phi_A)|\epsilon|^2+\frac12\int\frac{\phi_A'}{r}\cdot\frac{m^2}{r^2}|\epsilon|^2\\ -\sum_{\substack{\psi_1,\dots,\psi_4\in\{w^\flat,\epsilon\}:\\ \#\{i:\psi_i=\epsilon\}\geq2}}[\frac14\mathcal{M}_{4,0}^{(A)}+\frac m2\mathcal{M}_{4,1}^{(A)}]+\frac18\sum_{\substack{\psi_1,\dots,\psi_6\in\{w^\flat,\epsilon\}:\\ \#\{i:\psi_i=\epsilon\}\geq2}}\mathcal{M}_6^{(A)}.\end{aligned}$$

We further reduce (6.64) to

$$\lambda^2\partial_s\mathcal{I}_A-b\eta\theta_\eta M[\epsilon]=2bE_{w^\flat}^{(A),(\mathrm{qd})}[\epsilon]+\mathrm{Err}. \tag{6.66}$$

Indeed, we observe

$$\Big|\frac{\lambda_s}{\lambda}(R_{u^b-w^b},\Lambda_A z^b)_r\Big| \lesssim A\alpha^*\lambda^2\|\epsilon\|_{\dot H^1_m}^2,$$
$$\Big|\frac{\lambda_s}{\lambda}(R_{u^b-w^b},(\Lambda-\Lambda_A)Q_b^{(\eta)})_r\Big| \lesssim \lambda(o_{A\to\infty}(1)+\lambda)\|\epsilon\|_{\dot H^1_m}^2 + \lambda^2\|\epsilon\|_{L^2}\|\epsilon\|_{\dot H^1_m},$$
$$\Big|\Big(\frac{\lambda_s}{\lambda}+b\Big)E_{w^b}^{(A),(\mathrm{qd})}[\epsilon]\Big| \lesssim \lambda^{\frac{3}{2}}\|\epsilon\|_{\dot H^1_m}^2,$$

using

$$\Big|\frac{\lambda_s}{\lambda}\Big| \lesssim \lambda, \quad |\Lambda_A z^b| \lesssim A(|\partial_r z^b|+|r^{-1}z^b|), \quad |(\Lambda-\Lambda_A)Q_b^{(\eta)}| \lesssim \mathbf{1}_{r\geq A}(r^{-3}+br^{-1}),$$

the modulation estimate (6.28), and weak bootstrap hypothesis (5.25).

We now treat the term $2E_{w^b}^{(A),(\mathrm{qd})}[\epsilon]$ to conclude the proof of (6.58). Recall the expression (6.65). We can replace all the occurences of w^b by Q. Indeed, we use the multilinear estimates (Lemma 2.2) with $\|w^b-Q\|_{L^2}\lesssim \alpha^*+\lambda|\log\lambda|^{\frac{1}{2}}$ to see that those terms only contribute up to errors $(\alpha^*+\lambda|\log\lambda|^{\frac{1}{2}})\|\epsilon\|_{\dot H^1_m}^2$. Moreover, we can eliminate terms containing at least three ϵ's up to errors $\|\epsilon\|_{L^2}\|\epsilon\|_{\dot H^1_m}^2$. Thus we have (recall the formula of $\mathcal{J}_A$ given in (6.59))

$$E_{w^b}^{(A),(\mathrm{qd})}[\epsilon] = \mathcal{J}_A - \frac{1}{8}\int(\Delta^2\phi_A)|\epsilon|^2 + O\big((\alpha^*+\|\epsilon\|_{L^2}+\lambda|\log\lambda|^{\frac{1}{2}})\|\epsilon\|_{\dot H^1_m}^2\big).$$

Substituting this into (6.66), we conclude (6.58). □

The quantity

$$-\frac{1}{8}\int(\Delta^2\phi_A)|\epsilon|^2$$

does not have a good sign. Moreover, a crude bound

$$\Big|\int(\Delta^2\phi_A)|\epsilon|^2\Big| \lesssim \Big|\int \mathbf{1}_{r\geq A}\frac{A}{r}\cdot\Big|\frac{\epsilon}{r}\Big|^2\Big| \lesssim \|r^{-1}\epsilon\|_{L^2}^2 \lesssim \|\epsilon\|_{\dot H^1_m}^2$$

does not suffice. In order to go beyond this borderline, we crucially use the bound

$$|\Delta^2\phi_A| \lesssim \mathbf{1}_{r\geq A}\frac{A}{r^3},$$

which should be better than $\frac{1}{r^2}$. We will use an average argument.

We finally define the *Lyapunov/virial functional* by

$$\mathcal{I} := \frac{2}{\log A}\int_{A^{1/2}}^{A}\mathcal{I}_{A'}\frac{dA'}{A'} = \lambda^{-2}\Big(E_{w^b}^{(\mathrm{qd})}[\epsilon] + \frac{2}{\log A}\int_{A^{1/2}}^{A} b\Phi_{A'}[\epsilon]\frac{dA'}{A'} + \tfrac{\eta\theta_\eta}{2}M[\epsilon]\Big).$$

PROPOSITION 6.24 (Lyapunov/virial functional). *We have*

$$\lambda^2\mathcal{I} - \tfrac{\eta\theta_\eta}{2}M[\epsilon] + O\big((\alpha^*+\lambda^{\frac{1}{2}})\|\epsilon\|_{\dot H^1_m}^2 + A\lambda\|\epsilon\|_{L^2}\|\epsilon\|_{\dot H^1_m}\big) \sim \|\epsilon\|_{\dot H^1_m}^2, \tag{6.67}$$

$$\lambda^2\partial_s\mathcal{I} \geq \mathrm{Err}. \tag{6.68}$$

PROOF. The coercivity (6.67) follows from (6.57) and an averaging process. We turn to show (6.68). Let

$$\mathcal{J} := \frac{2}{\log A}\int_{A^{1/2}}^{A}\mathcal{J}_{A'}\frac{dA'}{A'}.$$

From the expression (6.58) and $b \lesssim \lambda$, it suffices to show

$$\frac{2}{\log A}\int_{A^{1/2}}^{A}\Big(\frac{1}{8}\int(\Delta^2\phi_A)|\epsilon|^2\Big)\frac{dA'}{A'} = o_{A\to\infty}(1)\|\epsilon\|_{\dot H^1_m}^2 \tag{6.69}$$

and

$$\mathcal{J} \geq o_{A\to\infty}(\|\epsilon\|_{\dot H^1_m}^2). \tag{6.70}$$

We first show (6.69). The crucial observations are

$$\frac{2}{\log A} = o_{A\to\infty}(1) \quad\text{and}\quad \int_{A^{1/2}}^{A}|\Delta^2\phi_{A'}|\frac{dA'}{A'} \lesssim \mathbf{1}_{r\geq A^{1/2}}\frac{1}{r^2}.$$

By Fubini's theorem, we see that

$$\frac{2}{\log A}\Big|\int_{A^{1/2}}^{A}\Big(\frac{1}{8}\int(\Delta^2\phi_{A'})|\epsilon|^2\Big)\frac{dA'}{A'}\Big| \lesssim o_{A\to\infty}(1)\|r^{-1}\epsilon\|_{L^2}^2.$$

This shows (6.69).

The rest of the proof is devoted to show the almost positivity (6.70). As there is a truncation on the region $\{r \leq A'\}$ in the expression (6.59) of $\mathcal{J}_{A'}$, it is natural to compare $\mathcal{J}_{A'}$ with a truncated version of $\frac{1}{2}(\mathcal{L}_Q\epsilon,\epsilon)_r = \frac{1}{2}\|L_Q\epsilon\|_{L^2}^2$ on the region $\{r \leq A'\}$. To realize this, we decompose the outermost integral $\int$ of $\mathcal{J}_{A'}$ as $\int \mathbf{1}_{r\leq A'} + \int \mathbf{1}_{r>A'} =: \mathcal{J}_{A'}^{(\leq)} + \mathcal{J}_{A'}^{(>)}$. If we can show that $\mathcal{J}_{A'}^{(\leq)} \approx \frac{1}{2}\|L_Q\epsilon\|_{L^2(r\leq A')}^2$ and $\mathcal{J}_{A'}^{(>)} \gtrsim 0$ in an averaged sense, then the almost positivity of $\mathcal{J}$ would follow.

For $\mathcal{J}_{A'}^{(\leq)}$, we can replace $\phi_{A'}(r)$ by $\frac{1}{2}r^2$ and expect that $\mathcal{J}_{A'}^{(\leq)}$ is close to $\frac{1}{2}\|L_Q\epsilon\|_{L^2(r\leq A')}^2$. We claim that

$$\frac{2}{\log A}\int_{A^{\frac{1}{2}}}^{A}\mathcal{J}_{A'}^{(\leq)}\frac{dA'}{A'} = \frac{1}{\log A}\int_{A^{\frac{1}{2}}}^{A}\Big(\int\mathbf{1}_{r\leq A'}|L_Q\epsilon|^2\Big)\frac{dA'}{A'} + o_{A\to\infty}(\|\epsilon\|_{\dot H^1_m}^2). \tag{6.71}$$

To see this, we observe by integration by parts that

$$\frac{1}{2}\Big(\int\mathbf{1}_{r\leq A'}|L_Q\epsilon|^2\Big) - \mathcal{J}_{A'}^{(\leq)} = 2\pi\cdot\Big(\frac{1}{2}(m+A_\theta[Q])|\epsilon|^2 - \epsilon\partial_r(Q\int_0^r \mathrm{Re}(Q\epsilon)r'dr')\Big)\Big|_{r=0}^{A'}.$$

Note that the boundary value at $r=0$ clearly vanishes. We can estimate the above as

$$\Big|\frac{1}{2}\Big(\int\mathbf{1}_{r\leq A'}|L_Q\epsilon|^2\Big) - \mathcal{J}_{A'}^{(\leq)}\Big| \lesssim |\epsilon(s,A')|^2 + o_{A'\to\infty}(\|\epsilon\|_{\dot H^1_m}^2).$$

Notice that we should not use a crude bound $|\epsilon(s,A')|^2 \lesssim \|\epsilon\|_{\dot H^1_m}^2$, which leads to an unacceptable error. Recalling that $\mathcal{J}$ is an *averaged* version of $\mathcal{J}_{A'}$ (this is also one of the reason why we need to average $\mathcal{J}_{A'}$), we obtain

$$\begin{aligned}&\frac{2}{\log A}\Big|\int_{A^{1/2}}^{A}\Big[\mathcal{J}_{A'}^{(\leq)} - \frac{1}{2}\Big(\int_0^{A'}\mathbf{1}_{r\leq A'}|L_Q\epsilon|^2\Big)\Big]\frac{dA'}{A'}\Big| \\ &\quad\lesssim \frac{2}{\log A}\int_{A^{1/2}}^{A}\frac{|\epsilon(s,A')|^2}{(A')^2}A'dA' + o_{A\to\infty}(\|\epsilon\|_{\dot H^1_m}^2) \lesssim o_{A\to\infty}(\|\epsilon\|_{\dot H^1_m}^2).\end{aligned}$$

This proves (6.71).

For $\mathcal{J}_{A'}^{(>)}$, we claim that

$$\frac{2}{\log A}\int_{A^{\frac{1}{2}}}^{A}\mathcal{J}_{A'}^{(>)}\frac{dA'}{A'} \geq o_{A\to\infty}(\|\epsilon\|_{\dot H^1_m}^2). \tag{6.72}$$

To see this, we use smallness of $\|Q\mathbf{1}_{r\geq A'}\|_{L^2} \lesssim (A')^{-(m+1)}$ to rearrange

$$\mathcal{J}_{A'}^{(>)} = \frac{1}{2}\int \mathbf{1}_{r\geq A'}\phi''_{A'}|\partial_r\epsilon|^2 + \frac{1}{2}\int \mathbf{1}_{r\geq A'}\frac{\phi'_{A'}}{r}\Big(\frac{m+A_\theta[Q]}{r}\Big)^2|\epsilon|^2 + o_{A'\to\infty}(\|\epsilon\|_{\dot H^1_m}^2).$$

Discarding nonnegative terms and averaging in A', we obtain (6.72).

Gathering the estimates (6.71) and (6.72), we obtain

$$\mathcal{J} \geq o_{A\to\infty}(\|\epsilon\|_{\dot H^1_m}^2).$$

This completes the proof of (6.70), and hence the proposition. □

6.6. Closing the bootstrap

Finally, we can finish the proof of our main bootstrap Lemma 5.3. The main tools are modulation estimates (Lemma 6.12), L^2-estimate of ϵ (Lemma 6.15), and Lyapunov/virial estimate (Proposition 6.24).

Proof of Lemma 5.3 (Finish). Applying (6.68) and the fundamental theorem of calculus with $\epsilon(0)=0$, we have

$$\lambda^2\mathcal{I}(t) \leq \lambda^2\int_t^0 \frac{1}{\lambda^2}(\partial_s\mathcal{I})dt' \lesssim \lambda^2\int_t^0 \frac{1}{\lambda^4}|\mathrm{Err}|dt'.$$

Applying (6.67), we obtain

$$\|\epsilon\|_{\dot H^1_m}^2 \lesssim (\alpha^*+\lambda^{\frac12})\|\epsilon\|_{\dot H^1_m}^2 + A\lambda\|\epsilon\|_{L^2}\|\epsilon\|_{\dot H^1_m} + \lambda^2\int_t^0 \frac{1}{\lambda^4}|\mathrm{Err}|dt'.$$

We then substitute into the above the L^2-bound (6.35), error bound (6.39), and maximal function estimate (2.12). But we need an extra care when we substitute $\lambda\cdot A(\alpha^*\lambda\eta)^2$ in place of Err. Indeed, we estimate it as

$$\begin{aligned}\lambda^2\int_t^0 \frac{1}{\lambda^4}\cdot\lambda A(\alpha^*\lambda\eta)^2dt' &= A\lambda^2\int_t^0 \frac{(\alpha^*\eta)^2}{\lambda}dt' \\ &\lesssim A\lambda^2\int_t^0 \frac{(\alpha^*\eta^{\frac34})^2}{\lambda^{\frac12}}dt' \lesssim A(\alpha^*\lambda^{\frac54}\eta^{\frac34})^2.\end{aligned}$$

Thus we obtain

$$\|\epsilon\|_{\dot H^1_m}^2 \lesssim A(\alpha^*\lambda^{m+2}+\alpha^*\lambda^{\frac54}\eta^{\frac34})^2 + (\alpha^*+o_{A\to\infty}(1)+A\lambda^{\frac14})(\mathcal{T}_{\dot H^1_m}^{(\eta,\frac54)}[\epsilon])^2.$$

Applying the time maximal function with (2.10) and (2.11), we get

$$\begin{aligned}\mathcal{T}_{\dot H^1_m}^{(\eta,\frac54)}[\epsilon] &\lesssim A^{\frac12}(\alpha^*\lambda^{m+2}+\alpha^*\lambda^{\frac54}\eta^{\frac34}) \\ &\quad + (\alpha^*+o_{A\to\infty}(1)+A\lambda^{\frac14})^{\frac12}\mathcal{T}_{\dot H^1_m}^{(\eta,\frac54)}[\epsilon].\end{aligned}$$

Note that $\mathcal{T}_{\dot H^1_m}^{(\eta,\frac54)}[\epsilon]$ has finite value.[6] If we choose A large such that $o_{A\to\infty}(1)\ll 1$ and t_0^* small such that $A\lambda^{\frac14} \lesssim A\langle t_0^*\rangle^{\frac14} \ll 1$, then for all sufficiently small α^* we have

$$\|\epsilon(t)\|_{\dot H^1_m} \leq \mathcal{T}_{\dot H^1_m}^{(\eta,\frac54)}[\epsilon] \lesssim \alpha^*\lambda^{m+2}+\alpha^*\lambda^{\frac54}\eta^{\frac34}.$$

Applying this strong bound into Lemma 6.35, we get

$$\|\epsilon(t)\|_{L^2} \lesssim \alpha^*\lambda^{m+\frac54}+\alpha^*\lambda^{\frac12}\eta^{\frac34}.$$

[6] As we are in the case $\eta>0$, $\mathcal{T}_{\dot H^1_m}^{(\eta,\frac54)}[\epsilon]$ is by definition finite.

We turn to estimate the size $|\lambda(t)-\langle t\rangle|$ and $|b(t)-|t||$. Applying the modulation estimates (Lemma 6.12), we have

$$\begin{aligned}\Big|\Big(\frac{\lambda^2}{b^2+\eta^2}\Big)_t\Big| &= \frac{2}{\lambda^2}\Big(\frac{\lambda^2}{b^2+\eta^2}\Big)\Big|\Big(\frac{\lambda_s}{\lambda}+b\Big)-\frac{(b_s+b^2+\eta^2)b}{b^2+\eta^2}\Big| \\ &\lesssim \alpha^*\lambda^m+\alpha^*\lambda^{-\frac{3}{4}}\eta^{\frac{3}{4}}.\end{aligned}$$

Integrating this from $t=0$, we get

$$\Big|\frac{\lambda^2}{b^2+\eta^2}-1\Big|\lesssim \alpha^*\lambda^{m+1}+\alpha^*\lambda^{\frac{1}{4}}\eta^{\frac{3}{4}}. \tag{6.73}$$

By (6.73) and modulation estimates (Lemma 6.12),

$$\begin{aligned}|b_t+1| &= \frac{1}{\lambda^2}\Big|(b_s+b^2+\eta^2)+(b^2+\eta^2)\Big(\frac{\lambda^2}{b^2+\eta^2}-1\Big)\Big| \\ &\lesssim \alpha^*\lambda^{m+1}+\alpha^*\lambda^{\frac{1}{4}}\eta^{\frac{3}{4}}.\end{aligned}$$

Integrating this from $t=0$, we have

$$\big|b-|t|\big|\lesssim \alpha^*\lambda^{m+2}+\alpha^*\lambda^{\frac{5}{4}}\eta^{\frac{3}{4}}.$$

Substituting this into (6.73), we get

$$|\lambda-\langle t\rangle| = \frac{|\lambda^2-(b^2+\eta^2)+(b^2-|t|^2)|}{\lambda+\langle t\rangle}\lesssim \alpha^*\lambda^{m+2}+\alpha^*\lambda^{\frac{5}{4}}\eta^{\frac{3}{4}}.$$

Finally, we estimate $\gamma(t)$. By (6.28),

$$\Big|\gamma_t-\frac{\eta\theta^{(\eta)}}{\lambda^2}+\theta_{z\to Q_b^{(\eta)\sharp}}\Big| = \frac{|\widetilde{\gamma}_s-\eta\theta^{(\eta)}|}{\lambda^2}\lesssim \alpha^*\lambda^m+\alpha^*\lambda^{-\frac{3}{4}}\eta^{\frac{3}{4}}.$$

Integrating this, we get

$$|\gamma-\gamma_\eta-\gamma_{\mathrm{cor}}^{(\eta)}|\lesssim \alpha^*\lambda^{m+1}+\alpha^*\lambda^{\frac{1}{4}}\eta^{\frac{3}{4}}.$$

This completes the proof of Lemma 5.3. □

REMARK 6.25. The use of $\mathcal{T}_{\dot{H}^1_m}^{(\eta,\frac{5}{4})}[\epsilon]$ is not necessary in the proof. One can indeed use $\mathcal{T}_{\dot{H}^1_m}^{(\eta,s)}[\epsilon]$ for any $s\in(1,\frac{5}{4}]$. The condition $s>1$ is used to have

$$\lambda^2\int_t^0\frac{1}{\lambda^4}\cdot\lambda(\mathcal{T}_{\dot{H}^1_m}^{(\eta,s)}[\epsilon])^2dt'\lesssim(\mathcal{T}_{\dot{H}^1_m}^{(\eta,s)}[\epsilon])^2$$

in $\lambda^2\int_t^0\frac{1}{\lambda^4}|\mathrm{Err}|dt'$. The condition $s\leq\frac{5}{4}$ is used to have

$$\mathcal{T}^{(\eta,s)}[\alpha^*\lambda^{\frac{5}{4}}\eta^{\frac{3}{4}}]\lesssim\alpha^*\lambda^{\frac{5}{4}}\eta^{\frac{3}{4}}.$$

REMARK 6.26. One can proceed the proof without using time maximal functions. In that way, however, one needs to assume stronger bootstrap hypothesis, for instance [**50**]. Using maximal functions, we can verify that weak bootstrap hypothesis suffices to propagate smallness of ϵ to the past times.

CHAPTER 7

Conditional uniqueness

This chapter is devoted to the proof of Theorem 1.2. As seen in the proof of Theorem 1.1, we have constructed blow-up solutions by a compactness argument. The (conditional) uniqueness of the constructed solutions was essential in the proof of Theorem 1.3, especially when we prove continuity of the map $\eta \in [0,\eta^*] \mapsto u^{(\eta)}$. Overall argument here is similar to as in Chapters 5 and 6. Given two solutions u_1 and u_2 satisfying the hypothesis of Theorem 1.2, we will decompose each u_j using their own modulation parameters and ϵ_j. After estimating differences of the modulation parameters, we show $\epsilon = \epsilon_1 - \epsilon_2 = 0$, by the Lyapunov/virial argument. Our approach is inspired by [**39**].[1]

We will decompose each $u_j(t)$ using modulation parameters $b_j(t)$, $\lambda_j(t)$, and $\gamma_j(t)$. We use the notation for modulated functions by

$$f_{b,\lambda,\gamma}(x) := \frac{1}{\lambda} f_b\Big(\frac{x}{\lambda}\Big) e^{i\gamma}$$

for $(b,\lambda,\gamma) \in \mathbb{R} \times \mathbb{R}_+ \times \mathbb{R}$.

Next, we will use $\sharp/\flat$ operations and dynamical rescalings. *Importantly, in the $\sharp/\flat$ operations, we only use $\lambda_1(t)$ and $\gamma_1(t)$.* Namely, if $f(s,y)$ and $g(t,x)$ are functions of (s,y) and (t,x), respectively, then we denote

$$\begin{aligned} f^\sharp(t,x) &= \frac{1}{\lambda_1(t)} f(s(t), \frac{x}{\lambda_1(t)}) e^{i\gamma_1(t)}, \\ f^\flat(s,y) &= \lambda_1(s) g(t(s), \lambda_1(s) y) e^{-i\gamma_1(s)}. \end{aligned}$$

The relation between (s,y) and (t,x) is given by

$$\frac{ds}{dt} := \frac{1}{\lambda_1^2} \quad \text{and} \quad y := \frac{x}{\lambda_1}. \tag{7.1}$$

Thus we will *not* use $\lambda_2(t)$ and $\gamma_2(t)$ when we rescale our solutions.

Finally, since we are in the $\eta = 0$ case, we will impose the *special relation* (5.20) between b_j and λ_j:

$$b_j(t) = |t|^{-1} (\lambda_j)^2(t). \tag{7.2}$$

The reasons for (7.2) are twofold. The first reason is as same as the dynamical law (5.18) with $\eta = 0$; the relation (7.2) cancels out the terms with the slowest

[1]There is an another strategy to prove the conditional uniqueness, used in [**50**, **57**] in the context of (NLS). There, they do not impose any orthogonality conditions on ϵ. Rather, they use a Lyapunov/virial argument to observe that ϵ can be controlled by some inner products, say $(\epsilon, \psi_b)_r$, where ψ is in the generalized nullspace of the linearized operator. On the other hand, one can control $(\epsilon, \psi_b)_r$ by a small constant times ϵ, by differentiating it in the s-variable several times. This argument yields $\epsilon = 0$. However, in case of (CSS), neither the quantities $(\epsilon, i\rho_b)_r$ nor $(\epsilon, |y|^2 Q_b)_r$ make sense if $m \in \{1,2\}$ due to the lack of decay of Q. Because of this, we could not follow their strategy.

spatial decay in the equation of ϵ_j. Secondly, (7.2) fixes the ratio $b_j/(\lambda_j)^2$ and makes $\|Q_{b_1,\lambda_1,0} - Q_{b_2,\lambda_2,0}\|_{L^2} \lesssim |\log(\frac{\lambda_1}{\lambda_2})|$ possible. One may compare this with the estimate $\|Q_b - Q\|_{L^2} \lesssim b|\log b|^{\frac{1}{2}}$, which is worse than (7.4) by a logarithmic factor. Having (7.4) is crucial in the proof of Theorem 1.2.[2]

LEMMA 7.1 (Estimate of difference). *Let $(b_j, \lambda_j, \gamma_j) \in \mathbb{R} \times \mathbb{R}_+ \times \mathbb{R}$ for $j \in \{1,2\}$ be such that*

$$\frac{b_1}{(\lambda_1)^2} = \frac{b_2}{(\lambda_2)^2} \tag{7.3}$$

and $|b_1| + |\frac{\lambda_1}{\lambda_2} - 1| + |\gamma_1 - \gamma_2| \le \frac{1}{2}$. Then, we have

$$\|\psi_{b_1} - \psi_{b_2, \frac{\lambda_2}{\lambda_1}, \gamma_2 - \gamma_1}\|_{H^1_m} \lesssim \Big(\Big|\log\Big(\frac{\lambda_1}{\lambda_2}\Big)\Big| + |\gamma_1 - \gamma_2| \Big), \qquad \forall \psi \in \{Q, \Lambda Q\}. \tag{7.4}$$

PROOF. Define the path $f : [0,1] \to H^1_m$ by

$$f(\tau) := \psi_{b_1(\lambda_1^{-\tau}\lambda_2^{\tau})^2, \lambda_1^{-\tau}\lambda_2^{\tau}, (\gamma_2 - \gamma_1)\tau}. \tag{7.5}$$

Note that $f(0) = \psi_{b_1}$ and $f(1) = \psi_{b_2, \frac{\lambda_2}{\lambda_1}, \gamma_2 - \gamma_1}$. Because of (7.3), we observe

$$\partial_\tau f = [-\log\Big(\frac{\lambda_2}{\lambda_1}\Big) \cdot \Lambda\psi + (\gamma_2 - \gamma_1) \cdot i\psi]_{b_1(\lambda_1^{-\tau}\lambda_2^{\tau})^2, \lambda_1^{-\tau}\lambda_2^{\tau}, (\gamma_2 - \gamma_1)\tau}.$$

By the fundamental theorem of calculus and Minkowski's inequality, the estimate (7.4) follows. □

7.1. A priori estimates on ϵ_1 and ϵ_2

Let

$$\theta_{\mathrm{cor}}(t) := -\int_0^\infty (-m - 2 + A_\theta[z(t)])|z(t)|^2 \frac{dr}{r}.$$

Note that θ_{cor} is equal to $\theta_{z \to Q_b^\sharp}$ in earlier sections. Recall also that

$$\gamma_{\mathrm{cor}}(t) = -\int_t^0 \theta_{\mathrm{cor}}(t')dt'.$$

LEMMA 7.2 (A priori estimates on ϵ_1 and ϵ_2). *Assume that two solutions u_1 and u_2 satisfy the hypothesis of Theorem 1.2 for some sufficiently small $c > 0$. For each $j \in \{1,2\}$, there exists unique $(b_j(t), \lambda_j(t), \gamma_j(t)) \in \mathbb{R} \times \mathbb{R}_+ \times \mathbb{R}$ for all t near 0 satisfying the following properties.*

(1) *(Decomposition) u_j has the decomposition*

$$u_j(t,x) = \frac{e^{i\gamma_j(t)}}{\lambda_j(t)} Q_{b_j(t)}\Big(\frac{x}{\lambda_j(t)}\Big) + z(t,x) + \frac{e^{i\gamma_1(t)}}{\lambda_1(t)} \epsilon_j(t, \frac{x}{\lambda_1(t)}) \tag{7.6}$$

with

$$\Big|\frac{\lambda_j}{|t|} - 1\Big| + |\gamma_j - \gamma_{\mathrm{cor}}| + \|\epsilon_j\|_{\dot H^1_m} + |t|\|\epsilon_j\|_{L^2_m} \lesssim c|t|^2, \tag{7.7}$$

$$(\epsilon_j, [\mathcal{Z}_{\mathrm{re}}]_{b_1})_r = (\epsilon_j, [i\mathcal{Z}_{\mathrm{im}}]_{b_1})_r = 0, \tag{7.8}$$

$$b_j = |t|^{-1}\lambda_j^2. \tag{7.9}$$

[2]Heuristically speaking, the situation is similar to when we prove uniqueness assertion of the contraction mapping principle. The uniqueness is guaranteed when the map is Lipschitz continuous with the Lipschitz constant less than 1, not when the map is merely Hölder continuous.

(2) *(Equation of ϵ_j) We have*

$$i\partial_t\epsilon_j^\sharp - \mathcal{L}_{w_j}\epsilon_j^\sharp = i\Big(\frac{(\lambda_j)_t}{\lambda_j} + |t|^{-1}\Big)[\Lambda Q]_{b_j,\lambda_j,\gamma_j} + ((\gamma_j)_t + \theta_{\mathrm{cor}})Q_{b_j,\lambda_j,\gamma_j} + \widetilde{R}_{Q_{b_j,\lambda_j,\gamma_j},z} + R_{u_j-w_j}. \tag{7.10}$$

(3) *(A priori modulation estimates) We have*

$$\Big|\frac{(\lambda_j)_s}{\lambda_j} + \Big(\frac{\lambda_1}{\lambda_j}\Big)^2 b_j\Big| + |(\gamma_j)_s + (\lambda_1)^2\theta_{\mathrm{cor}}| \lesssim (\alpha^* + c)(\lambda_1)^2. \tag{7.11}$$

REMARK 7.3. If we take the $\flat$-operation to (7.6) using λ_1 and γ_1, then we can write

$$u_1^\flat - u_2^\flat = [Q_{b_1} - Q_{b_2,\frac{\lambda_2}{\lambda_1},\gamma_2-\gamma_1}] + \epsilon_1 - \epsilon_2.$$

Here, we do not see modulation errors from z. Notice also that ϵ_1 and ϵ_2 are in the same scale, so we can study $\epsilon = \epsilon_1 - \epsilon_2$ without modulation errors.

PROOF. The decomposition of u_j will follow from the implicit function theorem. Let

$$\widetilde{u}_j(t,y) := |t|[u-z](t,|t|y)e^{-i\gamma_{\mathrm{cor}}(t)}.$$

We then have

$$\|\widetilde{u}_j(t) - Q_{|t|}\|_{\dot H^1_m} \le c|t|^2 \quad \text{and} \quad \|\widetilde{u}_j(t) - Q_{|t|}\|_{L^2} \le c|t| \tag{7.12}$$

for all t near 0.

We first consider the case $j=1$. For all t near 0, define the function $F_1^{(t)}$: $\dot H^1_m \times \mathbb{R} \times \mathbb{R}_+ \times \mathbb{R} \to \mathbb{R}^3$ by

$$F^{(t)}(v,b,\lambda,\gamma) := \begin{pmatrix} (v,[\mathcal{Z}_{\mathrm{re}}]_{b,\lambda,\gamma})_r - (Q,\mathcal{Z}_{\mathrm{re}})_r \\ (v,[i\mathcal{Z}_{\mathrm{im}}]_{b,\lambda,\gamma})_r \\ b - |t|\lambda^2 \end{pmatrix}.$$

Because of $\mathcal{Z}_{\mathrm{re}}, \mathcal{Z}_{\mathrm{im}} \in C^\infty_{c,m}(0,\infty)$, we see that $F_1^{(t)}$ is well-defined and in fact smooth. Note that

$$\frac{\delta F_1^{(t)}}{\delta v} = \begin{pmatrix} [\mathcal{Z}_{\mathrm{re}}]_{b,\lambda,\gamma} \\ [i\mathcal{Z}_{\mathrm{im}}]_{b,\lambda,\gamma} \\ 0 \end{pmatrix} = \begin{pmatrix} \mathcal{Z}_{\mathrm{re}} \\ i\mathcal{Z}_{\mathrm{im}} \\ 0 \end{pmatrix} + O_{(\dot H^1_m)^*}(|b| + |\lambda - 1| + |\gamma|)$$

and

$$\frac{\partial F_1^{(t)}}{\partial(b,\lambda,\gamma)} = \begin{pmatrix} 0 & -(\Lambda Q,\mathcal{Z}_{\mathrm{re}})_r & 0 \\ (Q,\frac{|y|^2}{4}\mathcal{Z}_{\mathrm{im}})_r & 0 & (Q,\mathcal{Z}_{\mathrm{im}})_r \\ 1 & 0 & 0 \end{pmatrix} + O(\|v - Q\|_{\dot H^1_m} + |t| + |b| + |\lambda - 1| + |\gamma|).$$

Finally, we notice $F_1^{(t)}(Q_{|t|},|t|,1,0) = 0$ and recall (7.12). By the above estimates, we can apply the implicit function theorem uniformly in t near 0 to have unique $(\widetilde{b}_1(t),\widetilde{\lambda}_1(t),\widetilde{\gamma}_1(t))$ such that

$$F_1^{(t)}(\widetilde{u}_1(t),\widetilde{b}_1(t),\widetilde{\lambda}_1(t),\widetilde{\gamma}_1(t)) = 0$$

and

$$|\widetilde{b}_1(t) - |t|| + |\widetilde{\lambda}_1(t) - 1| + |\widetilde{\gamma}_1(t)| \lesssim c|t|^2. \tag{7.13}$$

We now define

$$(b_1,\lambda_1,\gamma_1):=(\widetilde{b}_1,|t|\widetilde{\lambda}_1,\widetilde{\gamma}_1+\gamma_{\mathrm{cor}})$$

and recall the definition of $\widetilde{u}_1$. The conditions (7.8) and (7.9) for $j=1$ are clearly satisfied. For the estimate (7.7), we note that

$$\begin{aligned}\|\epsilon\|_{\dot H^1_m}&=\|[\widetilde{u}_1]_{\widetilde{\lambda}_1^{-1}(t),-\widetilde{\gamma}_1(t)}-Q_{b_1(t)}\|_{\dot H^1_m}\\&\lesssim\|\widetilde{u}_1-Q_{|t|}\|_{\dot H^1_m}+\|Q_{|t|}-Q_{b_1(t),\widetilde{\lambda}_1(t),\widetilde{\gamma}_1(t)}\|_{\dot H^1_m}\lesssim c|t|^2.\end{aligned}$$

Here the last inequality exploits (7.4). The L^2-estimate can be done similarly. This shows (7.7) for $j=1$.

We now consider the case $j=2$. For all t near 0, define the function $F_2^{(t)}$: $\dot H^1_m\times\mathbb{R}\times\mathbb{R}_+\times\mathbb{R}\to\mathbb{R}^3$ by

$$F_2^{(t)}(v,b,\lambda,\gamma):=\begin{pmatrix}(v-Q_{b,\lambda,\gamma},[\mathcal{Z}_{\mathrm{re}}]_{b_1(t),\widetilde{\lambda}_1(t),\widetilde{\gamma}_1(t)})_r\\(v-Q_{b,\lambda,\gamma},[i\mathcal{Z}_{\mathrm{im}}]_{b_1(t),\widetilde{\lambda}_1(t),\widetilde{\gamma}_1(t)})_r\\b-|t|\lambda^2\end{pmatrix}.$$

Because of $\mathcal{Z}_{\mathrm{re}},\mathcal{Z}_{\mathrm{im}}\in C^\infty_{c,m}(0,\infty)$, we see that $F_2^{(t)}$ is smooth. We note that

$$\frac{\delta F_2^{(t)}}{\delta v}=\begin{pmatrix}[\mathcal{Z}_{\mathrm{re}}]_{b_1(t),\widetilde{\lambda}_1(t),\widetilde{\gamma}_1(t)}\\ [i\mathcal{Z}_{\mathrm{im}}]_{b_1(t),\widetilde{\lambda}_1(t),\widetilde{\gamma}_1(t)}\\0\end{pmatrix}=\begin{pmatrix}\mathcal{Z}_{\mathrm{re}}\\i\mathcal{Z}_{\mathrm{im}}\\0\end{pmatrix}+O_{(\dot H^1_m)^*}(c|t|^2)$$

and

$$\begin{aligned}\frac{\partial F_2^{(t)}}{\partial(b,\lambda,\gamma)}&=\begin{pmatrix}0&-(\Lambda Q,\mathcal{Z}_{\mathrm{re}})_r&0\\(Q,\frac{|y|^2}{4}\mathcal{Z}_{\mathrm{im}})_r&0&(Q,\mathcal{Z}_{\mathrm{im}})_r\\1&0&0\end{pmatrix}\\&\quad+O(\|v-Q\|_{\dot H^1_m}+|t|+|b|+|\lambda-1|+|\gamma|).\end{aligned}$$

Notice that $F_2^{(t)}(Q_{b_1(t),\widetilde{\lambda}_1(t),\widetilde{\gamma}_1(t)},b_1(t),\widetilde{\lambda}_1(t),\widetilde{\gamma}_1(t))=0$. Applying the implicit function theorem, there exists unique $(\widetilde{b}_2(t),\widetilde{\lambda}_2(t),\widetilde{\gamma}_2(t))$ such that

$$F_1^{(t)}(\widetilde{u}_2(t),\widetilde{b}_2(t),\widetilde{\lambda}_2(t),\widetilde{\gamma}_2(t))=0$$

and

$$|\widetilde{b}_2(t)-b_1(t)|+|\widetilde{\lambda}_2(t)-\widetilde{\lambda}_1(t)|+|\widetilde{\gamma}_2(t)-\gamma_2(t)|\lesssim c|t|^2.$$

If we define

$$(b_2,\lambda_2,\gamma_2):=(\widetilde{b}_2,|t|\widetilde{\lambda}_2,\widetilde{\gamma}_{\mathrm{cor}}+\widetilde{\gamma}_2),$$

apply (7.13), recall the definition of $\widetilde{u}_2$, and proceed as in the case $j=1$, then (7.7)-(7.9) for $j=2$ follows. This completes the proof of the decomposition part of our lemma.

The equation (7.10) of $\epsilon_j^\sharp$ is clear from (6.26) with $\eta=0$. We turn to the modulation estimate (7.11). We take the $\flat$-operation to (7.10) and get

$$\begin{aligned}(7.14)\quad & i\partial_s\epsilon_j-\mathcal{L}_{w_j^\flat}\epsilon_j+ib_1\Lambda\epsilon_j\\&=i\Big(\frac{(\lambda_j)_s}{\lambda_j}+\frac{(\lambda_1)^2}{(\lambda_j)^2}b_j\Big)[\Lambda Q]_{b_j,\frac{\lambda_j}{\lambda_1},\gamma_j-\gamma_1}+((\gamma_j)_s+(\lambda_1)^2\theta_{\mathrm{cor}})Q_{b_j,\frac{\lambda_j}{\lambda_1},\gamma_j-\gamma_1}\\&\quad+i\Big(\frac{(\lambda_1)_s}{\lambda_1}+b_1\Big)\Lambda\epsilon_j+(\gamma_1)_s\epsilon_j+\widetilde{R}_{Q_{b_j,\frac{\lambda_j}{\lambda_1},\gamma_j-\gamma_1},z^\flat}+R_{u_j^\flat-w_j^\flat}.\end{aligned}$$

We then take the inner product of (7.14) with $i\psi_{b_1}$ for $\psi \in \{\mathcal{Z}_{\mathrm{re}}, i\mathcal{Z}_{\mathrm{im}}\}$ and use the orthogonality condition $(\epsilon, \psi_{b_1})_r = 0$. In case of $j = 1$, by the computation as in Lemma 6.12, we get

$$\Big|\frac{(\lambda_1)_s}{\lambda_1} + b_1\Big| + |(\gamma_1)_s + (\lambda_1)^2\theta_{\mathrm{cor}}| \lesssim \alpha^*(\lambda_1)^{m+3}|\log\lambda_1| + \|\epsilon_1\|_{\dot H_1^m} \lesssim (\alpha^* + c)(\lambda_1)^2.$$

Because of (7.2), we get

$$|(b_1)_s + (b_1)^2| \lesssim c(\lambda_1)^3.$$

Thus we have shown (7.11) for $j = 1$. In case of $j = 2$, we get

$$\begin{aligned}
&\Big(\frac{(\lambda_2)_s}{\lambda_2} + \frac{(\lambda_1)^2}{(\lambda_2)^2}b_2\Big)([\Lambda Q]_{b_2, \frac{\lambda_2}{\lambda_1}, \gamma_2-\gamma_1}, \psi_{b_1})_r + ((\gamma_2)_s + (\lambda_1)^2\theta_{\mathrm{cor}})(Q_{b_2, \frac{\lambda_2}{\lambda_1}, \gamma_2-\gamma_1}, i\psi_{b_1})_r \\
&= -(\epsilon_2, \mathcal{L}_{w_2^\flat} i\psi_{b_1})_r + \frac{(\lambda_1)_s}{\lambda_1}(\epsilon_2, \Lambda\psi_{b_1})_r - (\gamma_1)_s(\epsilon_2, i\psi_{b_1})_r + (b_1)_s(\epsilon_2, i\tfrac{|y|^2}{4}\psi_{b_1})_r \\
&\quad + (\widetilde{R}_{Q_{b_2, \frac{\lambda_2}{\lambda_1}, \gamma_2-\gamma_1}, z^\flat}, i\psi_{b_1})_r + (R_{u_2^\flat - w_2^\flat}, i\psi_{b_1})_r.
\end{aligned}$$

Substituting the modulation estimates for $(b_1, \lambda_1, \gamma_1)$ and estimating all the inner products by $\|\epsilon_2\|_{\dot H_m^1} + \alpha^*(\lambda_1)^{m+3}|\log\lambda_1|$, we get (7.11) for $j = 2$. ☐

7.2. Estimates of ϵ

Let

$$\epsilon := \epsilon_1 - \epsilon_2.$$

The main goal of this section is to derive the equation of ϵ and estimate error terms. The strategy is similar to that in Sections 6.1-6.4. Main issue here is to estimate the difference of two modulated profiles.

By (7.10), the difference $\epsilon = \epsilon_1 - \epsilon_2$ satisfies

$$\begin{aligned}
&i\partial_t\epsilon^\sharp - \mathcal{L}_{Q_{b_1,\lambda_1,\gamma_1}}\epsilon^\sharp \\
&= i\Big[\log\Big(\frac{\lambda_1}{\lambda_2}\Big)\Big]_t [\Lambda Q]_{b_1,\lambda_1,\gamma_1} + [\gamma_1 - \gamma_2]_t Q_{b_1,\lambda_1,\gamma_1} + (\lambda_1)^{-2}\Gamma^\sharp,
\end{aligned} \tag{7.15}$$

where

$$\begin{aligned}
(\lambda_1)^{-2}\Gamma^\sharp := {}& i\Big(\frac{(\lambda_2)_t}{\lambda_2} + |t|^{-1}\Big)([\Lambda Q]_{b_1,\lambda_1,\gamma_1} - [\Lambda Q]_{b_2,\lambda_2,\gamma_2}) \\
&+ ((\gamma_j)_t + \theta_{\mathrm{cor}})(Q_{b_1,\lambda_1,\gamma_1} - Q_{b_2,\lambda_2,\gamma_2}) + (\mathcal{L}_{w_1} - \mathcal{L}_{Q_{b_1,\lambda_1,\gamma_1}})\epsilon^\sharp \\
&+ (\mathcal{L}_{w_1} - \mathcal{L}_{w_2})\epsilon_2^\sharp \\
&+ (R_{u_1-w_1} - R_{u_2-w_2}) + (\widetilde{R}_{Q_{b_1,\lambda_1,\gamma_1}, z} - \widetilde{R}_{Q_{b_2,\lambda_2,\gamma_2}, z}).
\end{aligned} \tag{7.16}$$

Applying the $\flat$ operation, we get

$$\begin{aligned}
&i\partial_s\epsilon - \mathcal{L}_{Q_{b_1}}\epsilon + ib_1\Lambda\epsilon \\
&= i\Big[\log\Big(\frac{\lambda_1}{\lambda_2}\Big)\Big]_s [\Lambda Q]_{b_1} + [\gamma_1 - \gamma_2]_s Q_{b_1} + i\Big(\frac{(\lambda_1)_s}{\lambda_1} + b_1\Big)\Lambda\epsilon + (\gamma_1)_s\epsilon + \Gamma,
\end{aligned} \tag{7.17}$$

where

(7.18)

$$\Gamma := i\Big(\frac{(\lambda_2)_s}{\lambda_2}+\frac{\lambda_1^2}{\lambda_2^2}b_2\Big)([\Lambda Q]_{b_1}-[\Lambda Q]_{b_2,\frac{\lambda_2}{\lambda_1},\gamma_2-\gamma_1})$$
$$+((\gamma_2)_s+(\lambda_1)^2\theta_{\mathrm{cor}})(Q_{b_1}-Q_{b_2,\frac{\lambda_2}{\lambda_1},\gamma_2-\gamma_1})+(\mathcal{L}_{w_1^\flat}-\mathcal{L}_{Q_{b_1}})\epsilon+(\mathcal{L}_{w_1^\flat}-\mathcal{L}_{w_2^\flat})\epsilon_2$$
$$+(R_{u_1^\flat-w_1^\flat}-R_{u_2^\flat-w_2^\flat})+(\widetilde{R}_{Q_{b_1},z^\flat}-\widetilde{R}_{Q_{b_2,\frac{\lambda_2}{\lambda_1},\gamma_2-\gamma_1},z^\flat}).$$

LEMMA 7.4 (Various estimates). *We have the following*

(1) *(H^1_m-estimate of the difference) We have*

$$\|\psi_{b_1}-\psi_{b_2,\frac{\lambda_2}{\lambda_1},\gamma_2-\gamma_1}\|_{H^1_m}\lesssim(\lambda_1)^{-1}\mathcal{T}^{(\frac{5}{4})}_{\dot H^1_m}[\epsilon],\qquad\forall\psi\in\{\Lambda Q,iQ\}. \tag{7.19}$$

(2) *(H^1_m-estimate of Γ) We have*

$$\|\Gamma\|_{H^1_m}\lesssim(\alpha^*+c)\lambda_1\mathcal{T}^{(\frac{5}{4})}_{\dot H^1_m}[\epsilon]. \tag{7.20}$$

(3) *(Modulation estimate) We have*

$$\Big|\Big[\log\Big(\frac{\lambda_1}{\lambda_2}\Big)\Big]_s\Big|+|[\gamma_1-\gamma_2]_s|\lesssim\mathcal{T}^{(\frac{5}{4})}_{\dot H^1_m}[\epsilon]. \tag{7.21}$$

PROOF. We first claim

$$\|\Gamma\|_{H^1_m}\lesssim(\alpha^*\lambda_1+c(\lambda_1)^2)\|\epsilon\|_{\dot H^1_m}+(\alpha^*+c)(\lambda_1)^2\Big(\Big|\log\Big(\frac{\lambda_1}{\lambda_2}\Big)\Big|+|\gamma_1-\gamma_2|\Big). \tag{7.22}$$

To see this, we show that each term of (7.18) can be estimated by the RHS of (7.22). The first two terms of (7.18) can be treated by (7.11) and (7.4). For the third term of (7.18), since

$$w_1^\flat=Q_{b_1}+z^\flat,$$

we can write

$$(\mathcal{L}_{w_1^\flat}-\mathcal{L}_{Q_{b_1}})\epsilon$$
$$=\sum_{\substack{\psi_1,\psi_2,\psi_3\in\{Q_{b_1},z^\flat,\epsilon\}.\\ \#\{j:\psi_j=\epsilon\}=1,\\ \#\{j:\psi_j=z^\flat\}\geq1,}}[\mathcal{N}_{3,0}+\mathcal{N}_{3,1}+\mathcal{N}_{3,2}]+\sum_{\substack{\psi_1,\dots,\psi_5\in\{Q_{b_1},z^\flat,\epsilon\},\\ \#\{j:\psi_j=\epsilon\}=1,\\ \#\{j:\psi_j=z^\flat\}\geq1,}}[\mathcal{N}_{5,1}+\mathcal{N}_{5,2}].$$

We then apply (2.28) by distributing two $\dot H^1_m$ norms to $z^\flat$ and ϵ to get

$$\|(\mathcal{L}_{Q_b}-\mathcal{L}_{w_1^\flat})\epsilon\|_{H^1_m}\lesssim\alpha^*\lambda_1\|\epsilon\|_{\dot H^1_m}.$$

For the fourth term of (7.18), we write

$$w_1^\flat=(Q_{b_1}-Q_{b_2,\frac{\lambda_2}{\lambda_1},\gamma_2-\gamma_1})+w_2^\flat$$

and proceed as above by distributing two $\dot H^1_m$ norms to $Q_{b_1}-Q_{b_2,\frac{\lambda_2}{\lambda_1},\gamma_2-\gamma_1}$ and ϵ. By (2.28), (7.4), and (7.7), we get

$$\|(\mathcal{L}_{w_1^\flat}-\mathcal{L}_{w_2^\flat})\epsilon_2\|_{H^1_m}\lesssim\|Q_{b_1}-Q_{b_2,\frac{\lambda_2}{\lambda_1},\gamma_2-\gamma_1}\|_{\dot H^1_m}\|\epsilon_2\|_{\dot H^1_m}$$
$$\lesssim c(\lambda_1)^2\Big(\Big|\log\Big(\frac{\lambda_1}{\lambda_2}\Big)\Big|+|\gamma_1-\gamma_2|\Big).$$

Next, we treat the fifth term of (7.18). We write

$$w_1^\flat = (Q_{b_1} - Q_{b_2,\frac{\lambda_2}{\lambda_1},\gamma_2-\gamma_1}) + w_2^\flat \quad \text{and} \quad \epsilon_1 = \epsilon + \epsilon_2.$$

We distribute two $\dot{H}_m^1$ norms to either the pair ϵ_1 and $Q_{b_1} - Q_{b_2,\frac{\lambda_2}{\lambda_1},\gamma_2-\gamma_1}$, or ϵ and ϵ_2. By (2.28), (7.4), and (7.7), we have

$$\begin{aligned} &\|R_{u_1^\flat - w_1^\flat} - R_{u_2^\flat - w_2^\flat}\|_{H_m^1} \\ &\lesssim (\|Q_{b_1} - Q_{b_2,\frac{\lambda_2}{\lambda_1},\gamma_2-\gamma_1}\|_{\dot{H}_m^1} + \|\epsilon\|_{\dot{H}_m^1})(\|\epsilon_1\|_{\dot{H}_m^1} + \|\epsilon_2\|_{\dot{H}_m^1}) \\ &\lesssim c(\lambda_1)^2\Big(\Big|\log\Big(\frac{\lambda_1}{\lambda_2}\Big)\Big| + |\gamma_1 - \gamma_2| + \|\epsilon\|_{\dot{H}_m^1}\Big). \end{aligned}$$

Finally, we treat the last term of (7.18). We write

$$\widetilde{R}_{Q_{b_1},z^\flat} - \widetilde{R}_{Q_{b_2,\frac{\lambda_2}{\lambda_1},\gamma_2-\gamma_1},z^\flat} = \int_0^1 \partial_\tau(\widetilde{R}_{f(\tau),z^\flat})d\tau,$$

where $f(\tau)$ is the path connecting Q_{b_1} and $Q_{b_2,\frac{\lambda_2}{\lambda_1},\gamma_2-\gamma_1}$ as in (7.5). Observe that

$$\begin{aligned} &\sup_{\tau\in[0,1]} |\partial_\tau f(\tau)| \\ &\quad\lesssim \Big(\Big|\log\Big(\frac{\lambda_1}{\lambda_2}\Big)\Big| + |\gamma_1 - \gamma_2|\Big)F_{m,m+2}, \\ &\sup_{\tau\in[0,1]} |\partial_r\partial_\tau f(\tau)| + |r^{-1}\partial_\tau f(\tau)| \\ &\quad\lesssim \Big(\Big|\log\Big(\frac{\lambda_1}{\lambda_2}\Big)\Big| + |\gamma_1 - \gamma_2|\Big)(F_{m-1,m+3} + \lambda_1 F_{m+1,m+1}). \end{aligned}$$

Recalling the proof of Lemma 6.2 and applying Minkowski's inequality, we get

$$\|\widetilde{R}_{Q_{b_1},z^\flat} - \widetilde{R}_{Q_{b_2,\frac{\lambda_2}{\lambda_1},\gamma_2-\gamma_1},z^\flat}\|_{H_m^1} \lesssim \Big(\Big|\log\Big(\frac{\lambda_1}{\lambda_2}\Big)\Big| + |\gamma_1 - \gamma_2|\Big)\cdot \alpha^*\lambda_1^{m+3}|\log\lambda_1|.$$

This proves the claim.

We now claim the following preliminary version of the modulation estimates:

$$\Big|\Big[\log\Big(\frac{\lambda_1}{\lambda_2}\Big)\Big]_s\Big| + |[\gamma_1 - \gamma_2]_s| \lesssim \|\epsilon\|_{\dot{H}_m^1} + \|\Gamma\|_{L^2}. \tag{7.23}$$

Indeed, we proceed as in the proof of Lemma 6.11 to compute

$$\begin{aligned} 0 = \partial_s(\epsilon,\psi_{b_1})_r = (\epsilon,[\mathcal{L}_Q i\psi]_{b_1})_r + \Big[\log\Big(\frac{\lambda_1}{\lambda_2}\Big)\Big]_s(\Lambda Q,\psi)_r + [\gamma_1-\gamma_2]_s(Q,i\psi)_r \\ - \Big(\frac{(\lambda_1)_s}{\lambda_1} + b_1\Big)(\epsilon,[\Lambda\psi]_{b_1})_r + (\gamma_1)_s(\epsilon,i\psi_{b_1})_r + (\Gamma,i\psi_{b_1})_r \end{aligned}$$

for all $\psi \in \{\mathcal{Z}_{\mathrm{re}}, i\mathcal{Z}_{\mathrm{im}}\}$. Using $|(\epsilon,[\mathcal{L}_Q i\psi]_{b_1})_r| \lesssim \|\epsilon\|_{\dot{H}_m^1}$ and (7.11), we conclude (7.23).

With (7.4), (7.22), and (7.23) in hand, we are now in position to conclude the proof. We use (7.23), (7.22), and the fundamental theorem of calculus to get

$$\begin{aligned} \Big|\log\Big(\frac{\lambda_1}{\lambda_2}\Big)\Big| + |\gamma_1-\gamma_2| &\lesssim \int_t^0 \frac{1}{\lambda_1^2}(\|\epsilon\|_{\dot{H}_m^1} + \|\Gamma\|_{L^2})dt' \\ &\lesssim \frac{1}{\lambda_1}\Big(\mathcal{T}_{\dot{H}_m^1}^{(\frac{5}{4})}[\epsilon] + c(\lambda_1)^2\mathcal{T}^{(\frac{1}{4})}\Big[\Big|\log\Big(\frac{\lambda_1}{\lambda_2}\Big)\Big| + |\gamma_1-\gamma_2|\Big]\Big). \end{aligned}$$

Taking $\mathcal{T}^{(\frac{1}{4})}$ to the above and using (2.10) and (2.11), we get

$$\mathcal{T}^{(\frac{1}{4})}\Big[\Big|\log\Big(\frac{\lambda_1}{\lambda_2}\Big)\Big| + |\gamma_1-\gamma_2|\Big] \lesssim \lambda_1^{-1}\mathcal{T}^{(\frac{5}{4})}_{\dot{H}^1_m}[\epsilon].$$

Substituting this into (7.4), (7.22), and (7.23) again, we obtain (7.19), (7.20), and (7.21), respectively. This completes the proof. □

We also have L^2 -estimate of ϵ.

LEMMA 7.5 (L^2-estimate of ϵ). *We have*

$$\|\epsilon\|_{L^2} \lesssim (\lambda_1)^{-\frac{3}{4}}\mathcal{T}^{(\frac{5}{4})}_{\dot{H}^1_m}[\epsilon]. \tag{7.24}$$

PROOF. We recall the equation of $\epsilon^\sharp$ in the following form

$$\begin{aligned} i\partial_t\epsilon^\sharp + \Delta_m\epsilon^\sharp = i\Big[\log\Big(\frac{\lambda_1}{\lambda_2}\Big)\Big]_t[\Lambda Q]_{b_1,\lambda_1,\gamma_1} + [\gamma_1-\gamma_2]_t Q_{b_1,\lambda_1,\gamma_1} \\ + (\mathcal{L}_{Q_{b_1,\lambda_1,\gamma_1}} + \Delta_m)\epsilon^\sharp + (\lambda_1)^{-2}\Gamma^\sharp. \end{aligned}$$

We integrate the flow backward in time. By the Strichartz estimates (Lemma 2.7),

$$\|\epsilon^\sharp\|_{L^\infty_{[t,0)}L^2_x} \lesssim (7.25) + (7.26) + (7.27),$$

where

$$\Big\|i\Big[\log\Big(\frac{\lambda_1}{\lambda_2}\Big)\Big]_t[\Lambda Q]_{b_1,\lambda_1,\gamma_1} + [\gamma_1-\gamma_2]_t Q_{b_1,\lambda_1,\gamma_1}\Big\|_{L^{\frac{4}{3}}_{[t,0),x}}, \tag{7.25}$$

$$\|(\mathcal{L}_{Q_{b_1,\lambda_1,\gamma_1}} + \Delta_m)\epsilon^\sharp\|_{L^{\frac{4}{3}}_{[t,0),x}}, \tag{7.26}$$

$$\|(\lambda_1)^{-2}\Gamma^\sharp\|_{L^1_{[t,0)}L^2_x}. \tag{7.27}$$

For (7.25), we recall

$$\|[\Lambda Q]_{b_1,\lambda_1,\gamma_1}\|_{L^{\frac{4}{3}}_x} + \|Q_{b_1,\lambda_1,\gamma_1}\|_{L^{\frac{4}{3}}_x} \lesssim (\lambda_1)^{\frac{1}{2}}.$$

Thus by (7.21) and (2.12), we have

$$(7.25) \lesssim (\lambda_1)^{-\frac{3}{4}}\mathcal{T}^{(\frac{5}{4})}_{\dot{H}^1_m}[\epsilon].$$

The term (7.26) is treated as in the proof of Lemma 6.35:

$$(7.26) \lesssim (\lambda_1)^{-\frac{3}{4}}\mathcal{T}^{(\frac{5}{4})}_{\dot{H}^1_m}[\epsilon].$$

The remaining term (7.27) is treated using (7.20) and (2.12) as

$$(7.27) \lesssim (\alpha^* + c)\mathcal{T}^{(\frac{5}{4})}_{\dot{H}^1_m}[\epsilon].$$

This completes the proof. □

7.3. Lyapunov/virial functional

So far, we have estimated modulation parameters and various errors appearing in the ϵ-equation (7.17) by the $\dot{H}^1_m$-norm of ϵ. From now on, we control ϵ by introducing the Lyapunov/virial functional, as similarly as in Section 6.5.

For $A > 1$ to be chosen large later, we define

$$\mathcal{I}_A := \lambda^{-2}(E^{(\mathrm{qd})}_{Q_{b_1}}[\epsilon] + b_1\Phi_A[\epsilon])$$

and its averaged version

$$\mathcal{I} := \frac{2}{\log A}\int_{A^{1/2}}^{A} \mathcal{I}_{A'}\frac{dA'}{A'}.$$

We will collect errors satisfying

$$|\mathrm{Err}| \lesssim \lambda_1(\alpha^* + c + o_{A\to\infty}(1) + A\lambda^{\frac14})(\mathcal{T}^{(\frac54)}_{\dot{H}^1_m}[\epsilon])^2. \tag{7.28}$$

The main proposition in this section is as follows.

PROPOSITION 7.6 (Lyapunov/virial estimate). *We have*

$$\lambda^2\mathcal{I} + O(\lambda|\log\lambda|^{\frac12}\|\epsilon\|^2_{\dot{H}^1_m} + A\lambda\|\epsilon\|_{L^2}\|\epsilon\|_{\dot{H}^1_m}) \sim \|\epsilon\|^2_{\dot{H}^1_m}, \tag{7.29}$$

$$\lambda^2\partial_s\mathcal{I} \geq \mathrm{Err}. \tag{7.30}$$

As similarly as in Section 6.5, we start by computing $\partial_s E^{(\mathrm{qd})}_{Q_{b_1}}[\epsilon]$.

LEMMA 7.7 (Quadratic parts from energy). *We have*

$$E^{(\mathrm{qd})}_{Q_{b_1}}[\epsilon] + O(\lambda_1|\log\lambda_1|^{\frac12}\|\epsilon\|^2_{\dot{H}^1_m}) \sim \|\epsilon\|^2_{\dot{H}^1_m} \tag{7.31}$$

and

$$\begin{aligned}\partial_s E^{(\mathrm{qd})}_{Q_b}[\epsilon] &= (\mathcal{L}_{Q_{b_1}}\epsilon, \Big[\log\Big(\frac{\lambda_1}{\lambda_2}\Big)\Big]_s [\Lambda Q]_{b_1} - i[\gamma_1-\gamma_2]_s Q_{b_1})_r \\ &\quad + (\mathcal{L}_{Q_{b_1}}\epsilon + R_{(Q_{b_1}+\epsilon)-Q_{b_1}}, \frac{(\lambda_1)_s}{\lambda_1}\Lambda\epsilon)_r + \mathrm{Err}.\end{aligned} \tag{7.32}$$

PROOF. The coercivity (7.31) follows from the proof of (6.40) using $\|Q_{b_1} - Q\|_{L^2} \lesssim \lambda_1|\log\lambda_1|^{\frac12}$.

We turn to (7.32). We compute

$$\partial_s(E^{(\mathrm{qd})}_{Q_{b_1}}[\epsilon] = (R_{(Q_{b_1}+\epsilon)-Q_{b_1}}, \partial_s Q_{b_1})_r + (\mathcal{L}_{Q_{b_1}}\epsilon + R_{(Q_{b_1}+\epsilon)-Q_{b_1}}, \partial_s\epsilon)_r. \tag{7.33}$$

We note that $R_{(Q_{b_1}+\epsilon)-Q_{b_1}}$ satisfies the analogous estimates of (6.20) and (6.21) by the same proof. Thus

$$|(R_{(Q_{b_1}+\epsilon)-Q_{b_1}}, \partial_s Q_{b_1})_r| \lesssim |(b_1)_s|(\|\epsilon\|_{L^2} + \|\epsilon\|_{\dot{H}^1_m})\|\epsilon\|_{\dot{H}^1_m}.$$

Applying $|(b_1)_s| \lesssim (\lambda_1)^2$ and (7.24), the first term of the RHS of (7.33) is absorbed into Err.

We now treat the second term of the RHS of (7.33). We compute

$$(\mathcal{L}_{Q_{b_1}}\epsilon + R_{(Q_{b_1}+\epsilon)-Q_{b_1}}, \partial_s\epsilon)_r = (7.34) + (7.35) + (7.36) + (7.37),$$

where

$$(\mathcal{L}_{Q_{b_1}}\epsilon+R_{(Q_{b_1}+\epsilon)-Q_{b_1}},-i\mathcal{L}_{Q_{b_1}}\epsilon)_r \tag{7.34}$$

$$(\mathcal{L}_{Q_{b_1}}\epsilon+R_{(Q_{b_1}+\epsilon)-Q_{b_1}},\Big[\log\Big(\frac{\lambda_1}{\lambda_2}\Big)\Big]_s[\Lambda Q]_{b_1}-i[\gamma_1-\gamma_2]_sQ_{b_1})_r \tag{7.35}$$

$$(\mathcal{L}_{Q_{b_1}}\epsilon+R_{(Q_{b_1}+\epsilon)-Q_{b_1}},\frac{(\lambda_1)_s}{\lambda_1}\Lambda\epsilon-i(\gamma_1)_s\epsilon)_r \tag{7.36}$$

$$(i\mathcal{L}_{Q_{b_1}}\epsilon+R_{(Q_{b_1}+\epsilon)-Q_{b_1}},-i\Gamma)_r. \tag{7.37}$$

Let us treat (7.34)-(7.37) term by term. We arrange the term (7.34) as

$$(7.34)=(\mathcal{L}_{Q_{b_1}}\epsilon+R_{(Q_{b_1}+\epsilon)-Q_{b_1}},iR_{(Q_{b_1}+\epsilon)-Q_{b_1}})_r$$

and estimate it as

$$|(7.34)|\lesssim\|\epsilon\|_{\dot H^1_m}\|R_{(Q_{b_1}+\epsilon)-Q_{b_1}}\|_{H^1_m}\lesssim\|\epsilon\|_{\dot H^1_m}^3\lesssim c(\lambda_1)^2\|\epsilon\|_{\dot H^1_m}^2.$$

For the term (7.35), we can discard the terms having $R_{(Q_{b_1}+\epsilon)-Q_{b_1}}$:

$$\begin{aligned}|(R_{(Q_{b_1}+\epsilon)-Q_{b_1}},\Big[\log\Big(\frac{\lambda_1}{\lambda_2}\Big)\Big]_s[\Lambda Q]_{b_1}-i[\gamma_1-\gamma_2]_sQ_{b_1})_r|&\\ \lesssim\|\epsilon\|_{\dot H^1_m}^2\cdot\mathcal{T}^{(\frac{5}{4})}_{\dot H^1_m}[\epsilon]\lesssim c(\lambda_1)^2(\mathcal{T}^{(\frac{5}{4})}_{\dot H^1_m}[\epsilon])^2.&\end{aligned}$$

For (7.36), we keep the scaling term that is an unbounded looking quantity. The phase term is absorbed into Err as

$$|(\mathcal{L}_{Q_{b_1}}\epsilon+R_{(Q_{b_1}+\epsilon)-Q_{b_1}},-i(\gamma_1)_s\epsilon)_r|\lesssim|(\gamma_1)_s|\|\epsilon\|_{\dot H^1_m}^2\lesssim(\lambda_1)^2\|\epsilon\|_{\dot H^1_m}^2.$$

Finally, we treat (7.37) as an error term using (7.20):

$$|(7.37)|\lesssim\|\epsilon\|_{\dot H^1_m}\|\Gamma\|_{H^1_m}\lesssim(\alpha^*+c)\lambda_1\|\epsilon\|_{\dot H^1_m}\mathcal{T}^{(\frac{5}{4})}_{\dot H^1_m}[\epsilon].$$

This completes the proof. □

LEMMA 7.8 (Estimate of $b_1\Phi_A[\epsilon]$). *We have*

$$|b_1\Phi_A[\epsilon]|\lesssim A\lambda_1\|\epsilon\|_{L^2}\|\epsilon\|_{\dot H^1_m} \tag{7.38}$$

$$\begin{aligned}|\partial_s(b_1\Phi_A[\epsilon])|=&-\frac{(\lambda_1)_s}{\lambda_1}(\mathcal{L}_{Q_{b_1}}\epsilon,\Lambda_A\epsilon)_r\\&+(-ib_1\Lambda_A\epsilon,\Big[\log\Big(\frac{\lambda_1}{\lambda_2}\Big)\Big]_s[\Lambda Q]_{b_1}-i[\gamma_1-\gamma_2]_sQ_{b_1})_r+\mathrm{Err}.\end{aligned} \tag{7.39}$$

PROOF. The estimate (7.38) is an easy consequence of the Cauchy-Schwarz. Henceforth, we focus on (7.39). We compute

$$\partial_s(b_1\Phi_A[\epsilon])=(b_1)_s\Phi_A[\epsilon]+b_1(\Lambda_A\epsilon,i\partial_s\epsilon)_r=(b_1\Lambda_A\epsilon,i\partial_s\epsilon)_r+\mathrm{Err},$$

where we used $|(b_1)_s|\lesssim(\lambda_1)^2$. We then substitue the equation (7.17) of ϵ into the above to obtain

$$\partial_s(b_1\Phi_A[\epsilon])=(7.40)+(7.41)+(7.42)+(7.43)+\mathrm{Err},$$

where

$$(b_1\Lambda_A\epsilon, \mathcal{L}_{Q_{b_1}}\epsilon)_r, \tag{7.40}$$

$$(b_1\Lambda_A\epsilon, i\Big[\log\Big(\frac{\lambda_1}{\lambda_2}\Big)\Big]_s[\Lambda Q]_{b_1} + [\gamma_1-\gamma_2]_s Q_{b_1})_r, \tag{7.41}$$

$$(b_1\Lambda_A\epsilon, i\frac{(\lambda_1)_s}{\lambda_1}\Lambda\epsilon + (\gamma_1)_s\epsilon)_r, \tag{7.42}$$

$$(b_1\Lambda_A\epsilon, \Gamma)_r. \tag{7.43}$$

We treat (7.40)-(7.43) term by term. Firstly, we claim that (7.40) and the first term of the RHS of (7.39) agree up to Err. Indeed, the difference is given by

$$\Big(\frac{(\lambda_1)_s}{\lambda_1} + b_1\Big)(\Lambda_A\epsilon, \mathcal{L}_{Q_{b_1}}\epsilon)_r$$

and is absorbed into Err by mimicking the treatment of (6.53). Next, we keep the term (7.41). We then observe that (7.42) is absorbed into Err by mimicking the treatment of (6.55). Finally, we view (7.43) as an error by

$$|(7.43)| \lesssim A(\lambda_1)^2\|\epsilon\|_{\dot H^1_m}\|\Gamma\|_{L^2} \lesssim A(\lambda_1)^2(\mathcal{T}^{(\frac{5}{4})}_{\dot H^1_m}[\epsilon])^2.$$

This completes the proof. □

Proof of Proposition 7.6. Combining Lemmas 7.7 and 7.8, the rest of the arguments proceed similarly as in the proof of Lemma 6.22 and Proposition 6.24. We omit the details. □

7.4. Proof of conditional uniqueness

We are now in position to conclude the proof of the conditional uniqueness.

Proof of Theorem 1.2 (Finish). By (7.29), (7.30), and the fundamental theorem of calculus, we have

$$\|\epsilon\|^2_{\dot H^1_m} \lesssim \lambda_1|\log\lambda_1|^{\frac{1}{2}}\|\epsilon\|^2_{\dot H^1_m} + A\lambda_1\|\epsilon\|_{L^2}\|\epsilon\|_{\dot H^1_m} + \lambda^2\int_t^0 \frac{1}{\lambda^4}|\mathrm{Err}|dt'.$$

By (7.24), (7.28), and (2.12), we get

$$\|\epsilon\|^2_{\dot H^1_m} \lesssim (\alpha^* + c + o_{A\to\infty}(1) + A\lambda_1^{\frac{1}{4}})(\mathcal{T}^{(\frac{5}{4})}_{\dot H^1_m}[\epsilon])^2.$$

Applying the time maximal function and using (2.10), we get

$$\mathcal{T}^{(\frac{5}{4})}_{\dot H^1_m}[\epsilon] \lesssim (\alpha^* + c + o_{A\to\infty}(1) + A\lambda_1^{\frac{1}{4}})^{\frac{1}{2}}\mathcal{T}^{(\frac{5}{4})}_{\dot H^1_m}[\epsilon].$$

Recall that a priori decay of ϵ guarantees finiteness of $\mathcal{T}$. Thus $\mathcal{T}=0$, saying that $\epsilon = 0$. By the modulation estimates, this also says $\lambda_1 = \lambda_2$ and $\gamma_1 = \gamma_2$. From the relation $b_j = |t|^{-1}\lambda_j^2$, we have $b_1 = b_2$. Therefore, $u_1 = u_2$. □

APPENDIX A

Equivariant Sobolev spaces

In this appendix, we record basic properties of equivariant functions on $\mathbb{R}^2$ and associated Sobolev spaces. More precisely, we want to state and prove Sobolev and Hardy's inequality for equivariant functions. We assume equivariant function spaces and associated estimates are fairly standard. However, we could not find a well-organized reference discussing basic properties of equivariant functions. So we include, for self-containedness, a discussion on equivariant functions and required estimates for this work.

A.1. Smooth equivariant functions

Smooth radial functions. Let $f : \mathbb{R}^d \to \mathbb{C}$ be a radial function; let $g : \mathbb{R} \to \mathbb{C}$ be defined by $g(x) := f(x, 0, \dots, 0)$. We then see that $f(x) = g(|x|)$ for all $x \in \mathbb{R}^d$ and g is even. Conversely, if we are given an even function $g : \mathbb{R} \to \mathbb{C}$, then we can form a radial function $f : \mathbb{R}^d \to \mathbb{C}$ by $f(x) := g(|x|)$. We shall often view g by its restriction on $\mathbb{R}_+$ without mentioning. We note that if f is determined up to almost everywhere equivalence, so is g and vice versa.

If f is a smooth radial function, it is clear that g is smooth. We then ask the converse: is f smooth provided that g is smooth? The answer is affirmative, but we need the following preliminary lemma.

LEMMA A.1 (Removable singularity at zero). *Let $g : \mathbb{R} \to \mathbb{C}$ be a C^{k+1} function for some $k \geq 0$. Then,*

$$h(x) := \begin{cases} \frac{g(x) - g(0)}{x} & \text{if } x \neq 0, \\ g'(0) & \text{if } x = 0, \end{cases}$$

is a C^k function with

$$h^{(j)}(0) = \frac{1}{j+1} g^{(j+1)}(0), \qquad \forall j = 0, \dots, k.$$

In particular, if g is smooth, then so is h.

PROOF. Fix $k \geq 0$. Subtracting g by its $(k+1)$-th Taylor polynomial $\sum_{j=0}^{k+1} \frac{g^{(j)}(0)}{j!} x^j$, we may assume that g satisfies $g^{(j)}(0) = 0$ for all $0 \leq j \leq k+1$. In particular, we have

$$g^{(j)}(x) = o_{x \to 0}(|x|^{k+1-j}), \qquad \forall 0 \leq j \leq k+1. \tag{A.1}$$

For $0 \leq \ell \leq k$, let P_ℓ be the statement that h is of C^ℓ, $h^{(\ell)}(0) = 0$, and $h^{(\ell)}(x) = o_{x \to 0}(|x|^{k-\ell})$. We proceed by induction on ℓ. The statement P_0 is obvious from (A.1). We now assume $P_{\ell-1}$ for some $1 \leq \ell \leq k$ and show P_ℓ. It is clear that

h is of C^ℓ on $\mathbb{R} \setminus \{0\}$ and by inductive hypothesis, of $C^{\ell-1}$ on $\mathbb{R}$. To show the remaining assertions, the trick is to use the relation

$$xh(x) = g(x), \qquad \forall x \in \mathbb{R}.$$

We differentiate this relation ℓ-times to get

$$xh^{(\ell)}(x) = g^{(\ell)}(x) - \ell h^{(\ell-1)}(x), \qquad \forall x \in \mathbb{R} \setminus \{0\}.$$

Applying (A.1) and the inductive hypothesis, the proof of the inductive step (and hence the lemma) is complete. □

LEMMA A.2 (Smooth radial functions). *Let $g : \mathbb{R} \to \mathbb{C}$ be an even function; define a radial function $f : \mathbb{R}^d \to \mathbb{C}$ via $f(x) := g(|x|)$ for each $x \in \mathbb{R}^d$. Then, f is smooth if and only if g is smooth.*

PROOF. The difficult part is "if direction," which we now show. We use induction on k; let P_k be the statement that if g is a smooth even function, then f is of C^k. The statement P_0 is obviously true. Assuming P_k for some $k \geq 0$, we show P_{k+1}. We observe for each $i = 1, \dots, d$ that

$$\partial_i f(x) = x_i h(x), \qquad h(x) := \begin{cases} \frac{g'(|x|)}{|x|} & \text{if } x \neq 0, \\ g''(0) & \text{if } x = 0. \end{cases}$$

The radial part of h is smooth by Lemma A.1. Applying P_k to h, we see that h is of C^k. Thus f is of C^{k+1}, completing the inductive step and hence the proof. □

Smooth equivariant functions. Let $m \in \mathbb{Z} \setminus \{0\}$; let f be m-equivariant, i.e. $f(x) = g(r)e^{im\theta}$ for each $x \in \mathbb{R}^2 \setminus \{0\}$ with the expression $x_1 + ix_2 = re^{i\theta}$. Let $g : \mathbb{R} \to \mathbb{C}$ be a restriction of f defined by $g(x) := f(x, 0)$. We note that g is odd (resp., even) if m is odd (resp., even), which we say as *g has right parity*. Conversely, if we are given $m \in \mathbb{Z}$ and a function $g : \mathbb{R} \to \mathbb{C}$ having right parity, we can form an m-equivariant function f on $\mathbb{R}^2 \setminus \{0\}$ via $f(x) := g(r)e^{im\theta}$. To extend f on $\mathbb{R}^2$, the value of f at the origin is important. Setting $f(0) \neq 0$ yields discontinuity of f at the origin, because of the factor $e^{im\theta}$. Thus smoothness of f forces f to be degenerate at the origin. In this section, we want to study how much degeneracy f should have. The following lemma answers this.

LEMMA A.3 (Smooth equivariant functions). *Let $m \in \mathbb{Z} \setminus \{0\}$; let $g : \mathbb{R} \to \mathbb{C}$ be a function having right parity. Define $f : \mathbb{R}^2 \to \mathbb{C}$ by $f(x) := g(r)e^{im\theta}$ for $x \in \mathbb{R}^2 \setminus \{0\}$ and $f(0) := g(0)$. Then, f is smooth if and only if g is smooth and*

$$g^{(k)}(0) = 0$$

for each k such that $0 \leq k \leq |m| - 1$ or $k - m$ is odd. Moreover, if this is the case, we can express

$$f(x) = h(x) \cdot \begin{cases} (x_1 + ix_2)^m & \text{if } m > 0, \\ (x_1 - ix_2)^m & \text{if } m < 0, \end{cases} \qquad \forall x \in \mathbb{R}^2$$

for some smooth radial function h on $\mathbb{R}^2$.

PROOF. We first show the "only if direction." Since f must be continuous at the origin, we should set $f(0) = g(0) = 0$. Since $g(x) = f(x, 0)$ for all $x \in \mathbb{R}$, we see that g is smooth and $g^{(k)}(0) = 0$ when $k - m$ is odd since $g^{(k)}(x)$ has wrong parity. Henceforth, we focus on showing $g^{(k)}(0) = 0$ for each $k \leq |m| - 1$.

Given $k \geq 0$, let P_k be the statement that if f is a smooth m-equivariant function with $|m| \geq k+1$, then $g^{(j)}(0) = 0$ for all $j = 0, \dots, k$. The case of $k = 0$ is obvious by the continuity issue at the origin. We assume P_k for some $k \geq 0$ and show P_{k+1}. Namely, we assume $|m| \geq k+2$, know $g^{(0)}(0) = \cdots = g^{(k)}(0) = 0$ by inductive hypothesis, and aim to show $g^{(k+1)}(0) = 0$. Since f is smooth at the origin, the following limits must exist and do not depend on θ:

$$\lim_{r\to 0}(\partial_1 + i\partial_2)^{k+1}[g(r)e^{im\theta}] \quad \text{and} \quad \lim_{r\to 0}(\partial_1 - i\partial_2)^{k+1}[g(r)e^{im\theta}].$$

Recall the formulae

$$\partial_1 + i\partial_2 = e^{i\theta}[\partial_r + \tfrac{i}{r}\partial_\theta]; \qquad \partial_1 - i\partial_2 = e^{-i\theta}[\partial_r - \tfrac{i}{r}\partial_\theta].$$

In case of $m \geq k+2$, we compute using inductive hypothesis $g^{(0)}(0) = \cdots = g^{(k)}(0) = 0$ and Lemma A.1:

$$\begin{aligned}\lim_{r\to 0}(\partial_1 - i\partial_2)^{k+1}[g(r)e^{im\theta}] &= e^{i(m-k-1)\theta}\lim_{r\to 0}(\partial_r + \tfrac{m-k}{r})\cdots(\partial_r + \tfrac{m}{r})g(r).\\ &= e^{i(m-k-1)\theta}\Big(\prod_{j=0}^{k}\frac{k+m+1-2j}{j+1}\Big)g^{(k+1)}(0).\end{aligned}$$

As this expression should not depend on θ, we observe that $g^{(k+1)}(0) = 0$. The case of $m \leq -k-2$ can be treated similarly by computing $(\partial_1 + i\partial_2)^{k+1}f(0)$ instead. This completes the proof of the inductive step and hence the only if direction.

We turn to show the "if direction." By Lemma A.1, the function $\widetilde{h} : \mathbb{R} \to \mathbb{C}$ defined by

$$\widetilde{h}(r) := \begin{cases} r^{-|m|}g(r) & \text{if } r \neq 0 \\ \frac{1}{|m|!}g^{(|m|)}(0) & \text{if } r = 0 \end{cases}$$

is smooth and even. By Lemma A.2, the radial function $h : \mathbb{R}^2 \to \mathbb{C}$ defined by $h(x) := \widetilde{h}(|x|)$ is smooth. Therefore,

$$f(x) = h(x) \cdot \begin{cases} (x_1 + ix_2)^m & \text{if } m > 0 \\ (x_1 - ix_2)^m & \text{if } m < 0 \end{cases}$$

is smooth. This completes the proof. □

A.2. Equivariant Sobolev spaces

Radial Sobolev spaces. Let $L^2_{\mathrm{rad}}(\mathbb{R}^d)$ be the set of radial $L^2(\mathbb{R}^d)$ functions. Thus any $f \in L^2_{\mathrm{rad}}(\mathbb{R}^d)$ has a unique (up to a.e.) expression $f(x) = g(r)$ for some measurable function $g : \mathbb{R}_+ \to \mathbb{C}$, with the relation $r = |x|$. We moreover have the unitary equivalence

$$\begin{aligned}\Phi : L^2_{\mathrm{rad}}(\mathbb{R}^d) &\to L^2(\mathbb{R}_+, c_d r^{d-1}dr)\\ f &\mapsto g,\end{aligned}$$

where c_d is the area of the unit sphere of $\mathbb{R}^d$. This in particular shows that $L^2_{\mathrm{rad}}(\mathbb{R}^d)$ is complete and hence a closed subspace of $L^2(\mathbb{R}^d)$. The Fourier transform of $L^2_{\mathrm{rad}}(\mathbb{R}^d)$ functions are again radial.

For $s \geq 0$, the radial Sobolev space $H^s_{\mathrm{rad}}(\mathbb{R}^d)$ is defined by

$$H^s_{\mathrm{rad}}(\mathbb{R}^d) := H^s(\mathbb{R}^d) \cap L^2_{\mathrm{rad}}(\mathbb{R}^d).$$

Inner products are inherited from $H^s(\mathbb{R}^d)$. The set $\mathcal{S}_{\mathrm{rad}}(\mathbb{R}^d)$ of radial Schwartz functions is dense in $H^s_{\mathrm{rad}}(\mathbb{R}^d)$. For $H^1(\mathbb{R}^d)$ functions, we can define

$$\partial_r f := \frac{1}{r}\sum_{j=1}^{d} x_j \partial_j f.$$

At least for smooth radial functions, we have

$$\Delta f = \partial_{rr} f + \frac{d-1}{r}\partial_r f.$$

We moreover have

$$\|\nabla f\|_{L^2(\mathbb{R}^d)} = \|\partial_r f\|_{L^2(\mathbb{R}^d)} = \|\partial_r f\|_{L^2(\mathbb{R}_+, c_d r^{d-1} dr)}, \qquad \forall f \in H^1_{\mathrm{rad}}(\mathbb{R}^d).$$

In case of $d = 1$, any $H^1(\mathbb{R})$ functions are continuous and bounded. However, when $d \geq 2$, a $H^1_{\mathrm{rad}}(\mathbb{R}^d)$ function need not be bounded or continuous.

To deal with equivariant functions in the next section, we need the space $C^\infty_{c,\mathrm{rad}}(\mathbb{R}^d \setminus \{0\})$, which is the set of smooth radial functions having compact support in $\mathbb{R}^d \setminus \{0\}$. There is a well-known Hardy's inequality when $d \geq 3$:

$$\|\partial_r f\|_{L^2} \geq \frac{d-2}{2}\Big\|\frac{f}{r}\Big\|_{L^2}, \qquad \forall f \in C^\infty_{c,\mathrm{rad}}(\mathbb{R}^2 \setminus \{0\}).$$

We conclude this section by noting the density theorem by $C^\infty_{c,\mathrm{rad}}(\mathbb{R}^d \setminus \{0\})$ functions:

LEMMA A.4 (Density of $C^\infty_{c,\mathrm{rad}}(\mathbb{R}^d \setminus \{0\})$).

(1) *If $0 \leq s \leq \frac{d}{2}$, then $C^\infty_{c,\mathrm{rad}}(\mathbb{R}^d \setminus \{0\})$ is dense in $H^s_{\mathrm{rad}}(\mathbb{R}^d)$.*
(2) *If $s > \frac{d}{2}$, then $C^\infty_{c,\mathrm{rad}}(\mathbb{R}^d \setminus \{0\})$ is not dense in $H^s_{\mathrm{rad}}(\mathbb{R}^d)$.*

PROOF. The second assertion easily follows from the Sobolev embedding $H^s(\mathbb{R}^d) \hookrightarrow L^\infty(\mathbb{R}^d)$. When $s \leq \frac{d}{2}$, a crucial observation is the *failure* of $H^{\frac{d}{2}} \hookrightarrow L^\infty$; we can choose a radial Schwartz function g such that $\|g\|_{L^\infty} = g(0) = 1$ but $\|g\|_{H^{\frac{d}{2}}}$ is arbitrarily small. Note that $\mathcal{S}_{\mathrm{rad}}(\mathbb{R}^d)$ is dense in $H^s_{\mathrm{rad}}(\mathbb{R}^d)$, thanks to the Fourier transform. Therefore, to show that $C^\infty_{c,\mathrm{rad}}(\mathbb{R}^d \setminus \{0\})$ is dense in $H^s_{\mathrm{rad}}(\mathbb{R}^d)$, it suffices to approximate radial Schwartz functions f *with* $f(0) = 0$ by $C^\infty_{c,\mathrm{rad}}(\mathbb{R}^d \setminus \{0\})$ functions. Henceforth, we let χ to be a smooth radial bump function on $\mathbb{R}^d$, $\chi_R := \chi(\frac{\cdot}{R})$, and show that $\|\chi_R f\|_{H^{\frac{d}{2}}} \to 0$ as $R \to 0$.

For any $k \in \mathbb{N}$ and $0 < R \ll 1$, we observe by Leibniz's rule that

$$\begin{aligned}\|\chi_R f\|_{H^k} &\lesssim R^{-k+\frac{d}{2}} \|f\|_{L^\infty_{|x|\lesssim R}} + \sum_{j=1}^{k} \|\nabla^{k-j}\chi_R\|_{L^{2+}} \|\nabla^j f\|_{L^{\infty-}_{|x|\lesssim R}} \\ &\lesssim R^{-k+\frac{d}{2}} \|f\|_{L^\infty_{|x|\lesssim R}} + R^{-k+\frac{d}{2}+1-} \sum_{j=1}^{k} \|\nabla^j f\|_{L^{\infty-}_{|x|\lesssim R}}.\end{aligned}$$

When d is even, we choose $k = \frac{d}{2}$ to get

$$\|\chi_R f\|_{H^{\frac{d}{2}}} \lesssim \|f\|_{L^\infty_{|x|\lesssim R}} + R^{1-} \sum_{j=1}^{k} \|\nabla^j f\|_{L^{\infty-}_{|x|\lesssim R}}.$$

From our assumptions $f \in \mathcal{S}_{\text{rad}}(\mathbb{R}^d)$ and $f(0) = 0$, we take $R \to 0$ to conclude. When d is odd, we observe

$$\|\chi_R f\|_{H^{\frac{d}{2}}} \lesssim \|\chi_R f\|_{H^{\frac{d-1}{2}}}^{\frac{1}{2}} \|\chi_R f\|_{H^{\frac{d+1}{2}}}^{\frac{1}{2}} \to 0$$

as $R \to 0$. Note that $H^{\frac{d+1}{2}}$ norm may diverge with a factor $R^{-\frac{1}{2}}$ but this is cancelled with the factor $R^{\frac{1}{2}}$ from $H^{\frac{d-1}{2}}$ norm. This completes the proof. $\square$

Equivariant Sobolev spaces. We turn to equivariant Sobolev spaces. Define L^2_m by the set of m-equivariant $L^2(\mathbb{R}^2)$ functions. Inner product of L^2_m is inherited from $L^2(\mathbb{R}^2)$. Any $f \in L^2_m$ has a unique (up to a.e.) expression $f(x) = g(r)e^{im\theta}$ for some measurable function $g : \mathbb{R}_+ \to \mathbb{C}$. We have unitary equivalence

$$\begin{aligned} \Phi : L^2_m(\mathbb{R}^2) &\to L^2(\mathbb{R}_+, 2\pi r dr) \\ f &\mapsto g. \end{aligned}$$

For $s \geq 0$, the m-equivariant Sobolev space H^s_m is defined as

$$H^s_m := H^s(\mathbb{R}^2) \cap L^2_m,$$

equipped with the inner product of $H^s(\mathbb{R}^2)$. We define $\mathcal{S}_m$ by the set of m-equivariant Schwartz functions. We define $C^\infty_{c,m}(\mathbb{R}^2 \setminus \{0\})$ by the set of smooth m-equivariant functions having compact support in $\mathbb{R}^2 \setminus \{0\}$. Note that the Laplacian Δ has expression

$$\Delta = \partial_{rr} + \frac{1}{r}\partial_r - \frac{m^2}{r^2}$$

for m-equivariant functions.

Note that m-equivariant functions can be viewed as radial functions in higher dimensions. Indeed, if we define

$$\begin{aligned} \Psi : L^2_m &\to L^2_{\text{rad}}(\mathbb{R}^{2|m|+2}) \\ f &\mapsto c_m r^{-m} f \end{aligned}$$

for some universal constant c_m, then Ψ is unitary and the following diagram commutes:

$$\begin{array}{ccc} \mathcal{S}_m \subset L^2_m & \xrightarrow{\ \Psi\ } & \mathcal{S}_{\text{rad}}(\mathbb{R}^{2|m|+2}) \subset L^2_{\text{rad}}(\mathbb{R}^{2|m|+2}) \\ \downarrow \Delta & & \downarrow \Delta \\ \mathcal{S}_m \subset L^2_m & \xrightarrow{\ \Psi\ } & \mathcal{S}_{\text{rad}}(\mathbb{R}^{2|m|+2}) \subset L^2_{\text{rad}}(\mathbb{R}^{2|m|+2}) \end{array}$$

We note that $\Psi(\mathcal{S}_m) = \mathcal{S}_{\text{rad}}(\mathbb{R}^{2|m|+2})$, thanks to Lemma A.3. This allows us to transfer the density theorem for radial Sobolev spaces to equivariant Sobolev spaces.

LEMMA A.5 (Density of $C^\infty_{c,m}(\mathbb{R}^2 \setminus \{0\})$).

(1) *If* $0 \leq s \leq |m| + 1$, *then* $C^\infty_{c,m}(\mathbb{R}^2 \setminus \{0\})$ *is dense in* H^s_m.

(2) *If* $s > |m| + 1$, *then* $C^\infty_{c,m}(\mathbb{R}^2 \setminus \{0\})$ *is not dense in* H^s_m.

PROOF. The idea is to transfer the radial results using the above commutative diagram. Indeed, for $f, g \in \mathcal{S}_m$, we have

$$\begin{aligned} \|f - g\|^2_{H^{|m|+1}_m(\mathbb{R}^2)} &= (f - g, (1 - \Delta)^{|m|+1}(f - g)) \\ &= (\Psi(f - g), (1 - \Delta)^{|m|+1}\Psi(f - g)) \\ &= \|\Psi f - \Psi g\|_{H^{|m|+1}_{\text{rad}}(\mathbb{R}^{2|m|+2})}. \end{aligned}$$

1. It suffices to approximate m-equivariant Schwartz functions by $C_{c,m}^\infty(\mathbb{R}^2\setminus\{0\})$ in $H_m^{|m|+1}$ topology. Because of the above diagram and Lemma A.4, we know that $\Psi f \in \mathcal{S}_{\mathrm{rad}}(\mathbb{R}^{2|m|+2})$ can be approximated by a $C_{c,\mathrm{rad}}^\infty(\mathbb{R}^{2|m|+2}\setminus\{0\})$ function, say h. We then set $g := \Psi^{-1}h$.

2. This also follows from the transference. We omit the proof. □

For m-equivariant functions with $m \neq 0$, we have a nontrivial endpoint estimate of Hardy or Sobolev type. In general, it is well-known that the Hardy inequality fails for $d = 2$ and Sobolev inequality for $d = 2$ fails, i.e. $\dot{H}^1(\mathbb{R}^2) \not\hookrightarrow L^\infty(\mathbb{R}^2)$.

LEMMA A.6 (Hardy-Sobolev inequality). *Let $m \in \mathbb{Z}\setminus\{0\}$. For any $f \in H_m^1$, we have*

$$\|r^{-1}f\|_{L^2} + \|f\|_{L^\infty} \lesssim \|\nabla f\|_{L^2}.$$

PROOF. By density, we may assume $f \in C_{c,m}^\infty(\mathbb{R}^2\setminus\{0\})$. From the formula of Δ, we have

$$\|\nabla f\|_{L^2}^2 = (f, -\Delta f) = \|\partial_r f\|_{L^2}^2 + m^2\|r^{-1}f\|_{L^2}^2 \geq m^2\|r^{-1}f\|_{L^2}^2.$$

Another nice feature of m-equivariant functions is that we have embedding $\dot{H}_m^1 \hookrightarrow L^\infty$. Indeed,

$$|f(r)|^2 \leq \int_0^\infty |f||\partial_r f|dr' \leq \|r^{-1}f\|_{L^2}\|\partial_r f\|_{L^2} \lesssim \|\nabla f\|_{L^2}^2.$$

This completes the proof. □

We can generalize Hardy's inequality as follows.

LEMMA A.7 (Generalized Hardy's inequality). *For any $f \in C_{c,m}^\infty(\mathbb{R}^2\setminus\{0\})$ and $0 \leq k \leq |m|$, we have*

$$\|r^{-k}f\|_{L^2} + \|r^{-(k-1)}\partial_r f\|_{L^2} + \cdots + \|\partial_r^k f\|_{L^2} \sim_{k,m} \|f\|_{\dot{H}_m^k}.$$

PROOF. For simplicity, we only consider the case $m > 0$. The case $m < 0$ can be treated in a similar way, which we omit the proof. We claim that the following set of operators

$$\{e^{i(k-2\ell)\theta}(\partial_1 - i\partial_2)^{k-\ell}(\partial_1 + i\partial_2)^\ell\}_{0\leq\ell\leq k}$$

acting on $C_{c,m}^\infty(\mathbb{R}^2\setminus\{0\})$ is linearly independent. Indeed, we observe for $0 \leq j, \ell \leq k$ that

$$\begin{aligned}(\partial_1 + i\partial_2)^\ell[r^{m+2j}e^{im\theta}] &= 0, \qquad \text{if } \ell \geq j+1,\\ e^{i(k-2\ell)\theta}(\partial_1 - i\partial_2)^{k-\ell}(\partial_1 + i\partial_2)^\ell[r^{m+2\ell}e^{im\theta}] &= a_{k,\ell}r^{m+2\ell-k}e^{im\theta},\end{aligned}$$

where

$$a_{k,\ell} := 2^k(\ell!)\prod_{n=0}^{k-\ell-1}(m+\ell-n).$$

As $0 \leq k \leq m$, we have $a_{k,\ell} \neq 0$ for each $0 \leq \ell \leq k$. Namely, acting $r^{m+2j}e^{im\theta}$ on our set of operators and taking $r = 1$, we obtain a nonsingular matrix

$$\begin{bmatrix} a_{k,0} & * & * & \cdots & * \\ 0 & a_{k,1} & * & \cdots & * \\ 0 & 0 & a_{k,2} & \cdots & * \\ \vdots & \vdots & \vdots & \ddots & \vdots \\ 0 & 0 & 0 & \cdots & a_{k,k} \end{bmatrix}.$$

This proves the claim.

The claim immediately implies

$$\dim_{\mathbb{C}} \operatorname{span}_{\mathbb{C}}\{e^{i(k-2\ell)\theta}(\partial_1 - i\partial_2)^{k-\ell}(\partial_1 + i\partial_2)^{\ell}\}_{0\leq\ell\leq k} = k+1.$$

In view of

$$\operatorname{span}_{\mathbb{C}}\{e^{i(k-2\ell)\theta}(\partial_1 - i\partial_2)^{k-\ell}(\partial_1 + i\partial_2)^{\ell}\}_{0\leq\ell\leq k} \subseteq \operatorname{span}_{\mathbb{C}}\{r^{-(k-\ell)}\partial_r^{\ell}\}_{0\leq\ell\leq k},$$

we conclude the equality. Thus

$$\sum_{0\leq\ell\leq k} \|r^{-(k-\ell)}\partial_r^{\ell} f\|_{L^2} \sim_{k,m} \sum_{0\leq\ell\leq k} \|(\partial_1 + i\partial_2)^{\ell}(\partial_1 - i\partial_2)^{k-\ell} f\|_{L^2} \sim_{k,m} \|f\|_{\dot{H}_m^k}.$$

This completes the proof. ☐

LEMMA A.8 (Degeneracy at the origin). *Assume $s > |m| + 1$. Then,*

$$\sup_{r>0} \sum_{\ell=0}^{|m|} |r^{-(|m|-\ell)}\partial_r^{\ell} f| \lesssim_{m,s} \|f\|_{H_m^s}.$$

PROOF. Let f be m-equivariant. From the proof of Lemma A.7, we know that $\operatorname{span}_{\mathbb{C}}\{e^{i(|m|-2\ell)\theta}(\partial_1 - i\partial_2)^{|m|-\ell}(\partial_1 + i\partial_2)^{\ell}\}_{0\leq\ell\leq|m|} = \operatorname{span}_{\mathbb{C}}\{r^{-(|m|-\ell)}\partial_r^{\ell}\}_{0\leq\ell\leq|m|}$. From the usual Sobolev embedding $H^{1+} \hookrightarrow L^{\infty}$,

$$\begin{aligned} \sup_{r>0} \sum_{\ell=0}^{|m|} |r^{-(|m|-\ell)}\partial_r^{\ell} f| &\lesssim_m \sum_{\ell=0}^{|m|} \|(\partial_1 - i\partial_2)^{|m|-\ell}(\partial_1 + i\partial_2)^{\ell} f\|_{L^{\infty}} \\ &\lesssim_{m,s} \sum_{\ell=0}^{|m|} \|(\partial_1 - i\partial_2)^{|m|-\ell}(\partial_1 + i\partial_2)^{\ell} f\|_{H^{s-m}} \lesssim_{m,s} \|f\|_{H^s}. \end{aligned}$$

This completes the proof. ☐

APPENDIX B

Equivariant local theory

In this chapter, we sketch the proof of Proposition 2.10. Arguments here are close to Liu-Smith [**36**, Section 2]. By a standard contraction argument, the proof reduces to show the nonlinear estimates (Proposition B.4). Recall the nonlinearity of (1.22)

$$\mathcal{N}(\phi)=\frac{2m}{r^2}A_\theta\phi+\frac{1}{r^2}A_\theta^2\phi+A_0\phi-g|\phi|^2\phi.$$

Decompose $A_0=A_0^{(1)}+A_0^{(2)}$ such that

$$A_0^{(1)}[\phi]:=-\int_r^\infty A_\theta[\phi]|\phi|^2\frac{dr'}{r'},$$
$$A_0^{(2)}[\phi]:=-m\int_r^\infty |\phi|^2\frac{dr'}{r'}.$$

We denote the bilinear form associated to A_θ by

$$A_\theta[\psi_1,\psi_2]:=-\frac{1}{2}\int_0^r \mathrm{Re}(\psi_1\overline{\psi_2})r'dr'.$$

We will use the following estimates:

$$\Big\|\int_0^r f(r')r'dr'\Big\|_{L^\infty}\lesssim\|f\|_{L^1}, \tag{B.1}$$

$$\Big\|\frac{1}{r^2}\int_0^r f(r')r'dr'\Big\|_{L^2}\lesssim\|f\|_{L^2}, \tag{B.2}$$

$$\Big\|\int_r^\infty f(r')\frac{dr'}{r'}\Big\|_{L^2}\lesssim\|f\|_{L^2}. \tag{B.3}$$

Note that (B.1) follow from the observation $2\pi rdr=dx$ on $\mathbb{R}^2$. The estimates (B.2) and (B.3) follows by a change of variables and an application of Minkowski's inequality.

We first review the nonlinear estimate in Liu-Smith [**36**].

LEMMA B.1 (L^2-critical nonlinear estimates). *We have*

$$\|\mathcal{N}(\phi)\|_{L_{t,x}^{\frac43}}\lesssim(|m|+|g|+\|\phi\|_{L_t^\infty L_x^2}^2)\|\phi\|_{L_{t,x}^4}^3.$$

PROOF. Observe

$$\Big\|\frac{2m}{r^2}A_\theta\phi+\frac{1}{r^2}A_\theta^2\phi\Big\|_{L_{t,x}^{\frac43}}\lesssim(|m|+\|A_\theta\|_{L_{t,x}^\infty})\Big\|\frac{1}{r^2}A_\theta\Big\|_{L_{t,x}^2}\|\phi\|_{L_{t,x}^4},$$
$$\|A_0\phi\|_{L_{t,x}^{\frac43}}\leq\|A_0\|_{L_{t,x}^2}\|\phi\|_{L_{t,x}^4},$$
$$\|g|\phi|^2\phi\|_{L_{t,x}^{\frac43}}\leq|g|\|\phi\|_{L_{t,x}^4}^3.$$

Thus, it suffices to show the estimates

$$\|A_\theta\|_{L^\infty_{t,x}} \lesssim \|\phi\|^2_{L^\infty_t L^2_x}, \tag{B.4}$$

$$\Big\|\frac{1}{r^2}A_\theta\Big\|_{L^2_{t,x}} \lesssim \|\phi\|^2_{L^4_{t,x}}, \tag{B.5}$$

$$\|A_0\|_{L^2_{t,x}} \lesssim (|m| + \|\phi\|^2_{L^\infty_t L^2_x})\|\phi\|^2_{L^4_{t,x}}. \tag{B.6}$$

The estimate (B.4) follows from (B.1). The estimate (B.5) follows from (B.2). The estimate (B.6) follows from (B.3) and (B.4):

$$\|A_0\|_{L^2_{t,x}} \lesssim (|m| + \|A_\theta\|_{L^\infty_{t,x}})\Big\|\int_r^\infty |\phi|^2(r')\frac{dr'}{r'}\Big\|_{L^2_{t,x}} \lesssim (|m| + \|\phi\|^2_{L^\infty_t L^2_x})\|\phi\|^2_{L^4_{t,x}}.$$

This completes the proof. □

One can obtain analogous estimate for $\mathcal{N}(\phi_1) - \mathcal{N}(\phi_2)$. This allows us to get equivariant L^2-Cauchy theory as in Proposition 2.9.

We turn to H^s_m-subcritical Cauchy theory for $s > 0$. Nonlinearities involved in A_θ and A_0 are not of product type. In view of the Biot-Savart law (1.16), there is no (low) $\times$ (low) $\to$ (high) interaction in A_x, and hence in A_θ. However, it is not obvious that there is no (low) $\times$ (low) $\to$ (high) interaction in $\frac{1}{r^2}A_\theta$. A key observation in Liu-Smith [**36**, Section 4] is that (low) $\times$ (low) $\to$ (high) interaction is forbidden. This is formulated in the following two frequency localization lemmas.

Lemma B.2 (Fourier transform of A_θ and $r^{-2}A_\theta$ [**36**, Lemma 4.1]). *We have*

$$\mathcal{F}[A_\theta](\rho) = -\frac{1}{2\rho}\partial_\rho \mathcal{F}[|\phi|^2], \qquad \forall \rho \neq 0,$$
$$\mathcal{F}\Big[\frac{1}{r^2}A_\theta\Big](\rho) = -\frac{1}{2}\int_\rho^\infty \frac{1}{\rho'}\mathcal{F}[|\phi|^2]d\rho'.$$

More generally,

$$\mathcal{F}[A_\theta[\psi_1,\psi_2]](\rho) = -\frac{1}{2\rho}\partial_\rho \mathcal{F}[\mathrm{Re}(\overline{\psi_1}\psi_2)], \qquad \forall \rho \neq 0,$$
$$\mathcal{F}\Big[\frac{1}{r^2}A_\theta[\psi_1,\psi_2]\Big](\rho) = -\frac{1}{2}\int_\rho^\infty \frac{1}{\rho'}\mathcal{F}[\mathrm{Re}(\overline{\psi_1}\psi_2)]d\rho'.$$

Lemma B.3 (Fourier transform of A_0 [**36**, Lemma 4.2 and proof of Lemma 4.3]). *Define*

$$Q_{12}(f,g) := \partial_1 f \partial_2 g - \partial_2 f \partial_1 g.$$

We have

$$\mathcal{F}[A_0^{(1)}] = \frac{1}{\rho}\partial_\rho \mathcal{F}\Big[\frac{1}{r^2}A_\theta|\phi|^2\Big],$$
$$A_0^{(2)} = \Delta^{-1}\mathrm{Im}Q_{12}(\overline{\phi},\phi).$$

More generally,

$$\mathcal{F}\Big[\int_r^\infty \Big(\int_0^{r'} \mathrm{Re}(\overline{\psi_1}\psi_2) r'' dr''\Big) \mathrm{Re}(\overline{\psi_3}\psi_4) \frac{dr'}{r'}\Big]$$
$$= -\frac{1}{\rho}\partial_\rho \mathcal{F}\Big[\frac{1}{r^2}\Big(\int_0^{r} \mathrm{Re}(\overline{\psi_1}\psi_2) r' dr'\Big) \mathrm{Re}(\overline{\psi_3}\psi_4)\Big],$$
$$\int_r^\infty m \cdot \mathrm{Re}(\overline{\psi_1}\psi_2) \frac{dr'}{r'} = \Delta^{-1} \mathrm{Im} Q_{12}(\overline{\psi_1}, \psi_2).$$

We use Lemmas B.2 and B.3 to show the nonlinear estimate required for Proposition 2.10. We use the Besov norm

$$\|f\|_{B^s_{r,2}} := \|2^{js} P_j f\|_{\ell^2_{j\geq 0} L^r_x},$$

where $\{P_j\}_{j\geq 0}$ are the Littlewood-Paley projectors on $|\xi| \lesssim 1$ if $j=0$ and $|\xi| \sim 2^j$ if $j \geq 1$.

Proposition B.4 (Nonlinear estimates; higher regularity). *Let $s>0$. Then,*

$$\|\mathcal{N}(\phi)\|_{L^{\frac{4}{3}}_t B^s_{4/3,2}} \lesssim_s (|m| + |g| + \|\phi\|^2_{L^\infty_t L^2_x}) \|\phi\|^2_{L^4_{t,x}} \|\phi\|_{L^4_t B^s_{4,2}}$$
$$+ \|\phi\|_{L^\infty_t L^2_x} \|\phi\|^3_{L^4_{t,x}} \|\phi\|_{L^\infty_t H^s_x}.$$

Proof. In what follows, implicit constants are allowed to depend on s. Let $\frac{1}{p} = \frac{1}{p_1} + \frac{1}{p_2}$ with $1 \leq p, p_1, p_2 \leq \infty$. We then have a standard paraproduct estimate

$$\Big\|2^{ks} \sum_{j\geq k} \|(P_{\leq j} f)(P_j g)\|_{L^p_x}\Big\|_{\ell^2_k} \lesssim \|P_{\leq j} f\|_{\ell^\infty_j L^{p_1}_x} \Big\|\sum_{j\geq k} 2^{(k-j)s} \cdot 2^{js} \|P_j g\|_{L^p_x}\Big\|_{\ell^2_k}$$
$$\lesssim \|f\|_{L^{p_1}_x} \|g\|_{B^s_{p_2,2}},$$

where we used Schur's test in the second inequality. In particular, we have

$$\|2^{js}(P_{\leq j} f)(P_j g)\|_{\ell^2_j L^p_x} \lesssim \|f\|_{L^{p_1}_x} \|g\|_{B^s_{p_2,2}},$$
$$\|fg\|_{B^s_{p,2}} \lesssim \|f\|_{L^{p_1}_x} \|g\|_{B^s_{p_2,2}} + \|f\|_{B^s_{p_1,2}} \|g\|_{L^{p_2}_x}.$$

Thus we can estimate each term of the nonlinearity as

$$\|g|\phi|^2\phi\|_{L^{\frac{4}{3}}_t B^s_{4/3,2}} \lesssim |g| \|\phi\|^2_{L^4_{t,x}} \|\phi\|_{L^4_t B^s_{4,2}},$$
$$\Big\|\frac{2m}{r^2} A_\theta \phi\Big\|_{L^{\frac{4}{3}}_t B^s_{4/3,2}} \lesssim |m| \Big(\Big\|\frac{1}{r^2} A_\theta\Big\|_{L^2_{t,x}} \|\phi\|_{L^4_t B^s_{4,2}} + \Big\|\frac{1}{r^2} A_\theta\Big\|_{L^2_t B^s_{2,2}} \|\phi\|_{L^4_{t,x}}\Big),$$
$$\Big\|\frac{1}{r^2} A_\theta^2 \phi\Big\|_{L^{\frac{4}{3}}_t B^s_{4/3,2}} \lesssim \Big(\Big\|\frac{1}{r^2} A_\theta\Big\|_{L^2_{t,x}} \|A_\theta\phi\|_{L^4_t B^s_{4,2}} + \Big\|\frac{1}{r^2} A_\theta\Big\|_{L^2_t B^s_{2,2}} \|A_\theta\phi\|_{L^4_{t,x}}\Big),$$
$$\|A_0\phi\|_{L^{\frac{4}{3}}_t B^s_{4/3,2}} \lesssim (\|A_0\|_{L^2_{t,x}} \|\phi\|_{L^4_t B^s_{4,2}} + \|A_0\|_{L^2_t B^s_{2,2}} \|\phi\|_{L^4_{t,x}}).$$

For the terms with L^2-critical Strichartz pairs, we can proceed as in Lemma B.1. Thus it suffices to show the estimates

$$\|A_\theta \phi\|_{L_t^4 B_{4,2}^s} \lesssim \|\phi\|_{L_t^\infty L_x^2} (\|\phi\|_{L_{t,x}^4} \|\phi\|_{L_t^\infty H_x^s} + \|\phi\|_{L_t^\infty L_x^2} \|\phi\|_{L_t^4 B_{4,2}^s}), \tag{B.7}$$

$$\Big\|\frac{1}{r^2} A_\theta\Big\|_{L_t^2 B_{2,2}^s} \lesssim \|\phi\|_{L_{t,x}^4} \|\phi\|_{L_t^4 B_{4,2}^s}, \tag{B.8}$$

$$\|A_0^{(1)}\|_{L_t^2 B_{2,2}^s} \lesssim \|\phi\|_{L_t^\infty L_x^2} \|\phi\|_{L_{t,x}^4} (\|\phi\|_{L_{t,x}^4} \|\phi\|_{L_t^\infty H_x^s} + \|\phi\|_{L_t^\infty L_x^2} \|\phi\|_{L_t^4 B_{4,2}^s}), \tag{B.9}$$

$$\|A_0^{(2)}\|_{L_t^2 B_{2,2}^s} \lesssim |m| \|\phi\|_{L_{t,x}^4} \|\phi\|_{L_t^4 B_{4,2}^s}. \tag{B.10}$$

Our main tools are (B.1)-(B.3) and Lemmas (B.2)-(B.3).

To show (B.7), we use Lemma B.2 to observe[1]

$$\begin{aligned} P_k(A_\theta \phi) &= P_k\Big(\Big(2 \sum_{\max\{j_1,j_2,j_3\}=j_1\geq k} + \sum_{\max\{j_1,j_2,j_3\}=j_3\geq k}\Big) A_\theta[P_{j_1}\phi, P_{j_2}\phi] P_{j_3}\phi\Big) \\ &= P_k\Big(2\sum_{j\geq k} A_\theta[P_j\phi, P_{\leq j}\phi] P_{\leq j}\phi + \sum_{j\geq k} A_\theta[P_{\leq j}\phi] P_j\phi\Big). \end{aligned}$$

We then estimate

$$\begin{aligned} \|A_\theta\phi\|_{L_t^4 B_{4,2}^s} &\lesssim \Big\|2^{sk}\sum_{j\geq k} \|A_\theta[P_j\phi, P_{\leq j}\phi]\|_{L_x^\infty} \|P_{\leq j}\phi\|_{L_x^4}\Big\|_{L_t^4 \ell_k^2} \\ &\quad + \Big\|2^{sk}\sum_{j\geq k} \|A_\theta[P_{\leq j}\phi]\|_{L_x^\infty} \|P_j\phi\|_{L_x^4}\Big\|_{L_t^4 \ell_k^2} \\ &\lesssim \|P_{\leq j}\phi\|_{L_t^\infty \ell_j^\infty L_x^2} \|P_{\leq j}\phi\|_{L_t^4 \ell_j^\infty L_x^4} \Big\|2^{sk}\sum_{j\geq k} P_j\phi\Big\|_{L_t^\infty \ell_k^2 L_x^2} \\ &\quad + \|P_{\leq j}\phi\|_{L_t^\infty \ell_j^\infty L_x^2}^2 \Big\|2^{sk}\sum_{j\geq k} P_j\phi\Big\|_{L_t^4 \ell_k^2 L_x^4} \\ &\lesssim \|\phi\|_{L_t^\infty L_x^2} (\|\phi\|_{L_{t,x}^4} \|\phi\|_{L_t^\infty H_x^s} + \|\phi\|_{L_t^\infty L_x^2} \|\phi\|_{L_t^4 B_{4,2}^s}). \end{aligned}$$

To show (B.8), we use Lemma B.2 to observe

$$P_k\Big(\frac{1}{r^2} A_\theta\Big) = P_k\Big(\frac{1}{r^2} \sum_{\max\{j,\ell\}\geq k} A_\theta[P_j\phi, P_\ell\phi]\Big) = 2P_k\Big(\frac{1}{r^2}\sum_{j\geq k} A_\theta[P_j\phi, P_{\leq j}\phi]\Big).$$

We then use (B.2) to estimate

$$\begin{aligned} \Big\|\frac{1}{r^2} A_\theta\Big\|_{L_t^2 B_2^{s,2}} &\lesssim \Big\|2^{sk} P_k\big(\frac{1}{r^2}\sum_{j\geq k} A_\theta[P_j\phi, P_{\leq j}\phi]\big)\Big\|_{L_t^2 \ell_k^2 L_x^2} \\ &\lesssim \Big\|2^{sk}\sum_{j\geq k} \|P_j\phi P_{\leq j}\phi\|_{L_x^2}\Big\|_{L_t^2 \ell_k^2} \lesssim \|\phi\|_{L_{t,x}^4} \|\phi\|_{L_t^4 B_{4,2}^s}. \end{aligned}$$

[1]The display is not exactly correct, but it captures essential features of the proof.

To show (B.9), we use Lemma B.3 to observe

$$
\begin{aligned}
P_k A_0^{(1)} &= 2P_k\Big(\Big(\sum_{\max\{j_1,\dots j_4\}=j_1\geq k} + \sum_{\max\{j_1,\dots,j_4\}=j_3\geq k}\Big) \\
&\qquad \times \int_r^\infty A_\theta[P_{j_1}\phi, P_{j_2}\phi]\mathrm{Re}(\overline{P_{j_3}\phi}P_{j_4}\phi)\frac{dr'}{r'}\Big) \\
&= 2P_k\Big(\sum_{j\geq k}\int_r^\infty \frac{1}{r'}A_\theta[P_j\phi, P_{\leq j}\phi]|P_{\leq j}\phi|^2 dr' \\
&\qquad + \sum_{j\geq k}\int_r^\infty A_\theta[P_{\leq j}\phi]\mathrm{Re}(\overline{P_j\phi}P_{\leq j}\phi)\frac{dr'}{r'}\Big).
\end{aligned}
$$

We then use (B.3) and (B.1) to estimate

$$
\begin{aligned}
\|A_0^{(1)}\|_{L_t^2 B_2^{s,2}} &\lesssim \Big\|2^{sk}\sum_{j\geq k}\|A_\theta[P_j\phi, P_{\leq j}\phi]\|_{L_x^\infty}\|P_{\leq j}\phi\|_{L_x^4}^2\Big\|_{L_t^2\ell_k^2} \\
&\quad + \Big\|2^{sk}\sum_{j\geq k}\|A_\theta[P_{\leq j}\phi]\|_{L_x^\infty}\|P_j\phi\|_{L_x^4}\|P_{\leq j}\phi\|_{L_x^4}\Big\|_{L_t^2\ell_k^2} \\
&\lesssim \|P_{\leq j}\phi\|_{L_t^\infty\ell_j^\infty L_x^2}\|P_{\leq j}\phi\|_{L_t^4\ell_j^\infty L_x^4}^2 \cdot \Big\|2^{sk}\sum_{j\geq k}\|P_j\phi\|_{L_x^2}\Big\|_{L_t^\infty\ell_k^2} \\
&\quad + \|P_{\leq j}\phi\|_{L_t^\infty\ell_j^\infty L_x^2}^2\|P_{\leq j}\phi\|_{L_t^4\ell_j^\infty L_x^4}\Big\|2^{sk}\sum_{j\geq k}\|P_j\phi\|_{L_x^4}\Big\|_{L_t^4\ell_k^2} \\
&\lesssim \|\phi\|_{L_t^\infty L_x^2}\|\phi\|_{L_{t,x}^4}^2\|\phi\|_{L_t^\infty H_x^s} + \|\phi\|_{L_t^\infty L_x^2}^2\|\phi\|_{L_{t,x}^4}\|\phi\|_{L_t^4 B_{4,2}^s}.
\end{aligned}
$$

To show (B.10), we use Lemma B.3 to observe

$$
P_k A_0^{(2)} = 2P_k\sum_{j\geq k}\int_r^\infty \frac{m}{r'}\mathrm{Re}(\overline{P_j\phi}P_{\leq j}\phi)dr'.
$$

We then use (B.3) to estimate

$$
\|A_0^{(2)}\|_{L_t^2 B_2^{s,2}} \lesssim |m|\Big\|2^{sk}\sum_{j\geq k}\|(P_j\phi)(P_{\leq j}\phi)\|_{L_x^4}\Big\|_{L_t^2\ell_k^2} \lesssim |m|\|\phi\|_{L_{t,x}^4}\|\phi\|_{L_t^4 B_{4,2}^s}.
$$

This completes the proof. □

One can get an analogous estimate for the difference $\mathcal{N}(\phi_1) - \mathcal{N}(\phi_2)$. One then has local existence and uniqueness for H_m^s data by the standard contraction principle. However, the local existence in *subcritical* sense is not clear. To get this, we use Hölder's inequality in time and the embedding $B_{4-,2}^s \hookrightarrow L^4$ to observe

$$
\|\phi\|_{L_{I,x}^4} \lesssim_s |I|^{0+}\|\phi\|_{L_{|I|}^{4+}B_{4-,2}^s}
$$

such that above pair $(4+, 4-)$ is admissible. Observe that in Proposition B.4, we have $L_{t,x}^4$ factor in the upper bound. This guarantees local existence in subcritical sense. Thus H_m^s-subcritical local theory (Proposition 2.10) follows by standard arguments.

Bibliography

[1] L. Bergé, A. De Bouard, and J.-C. Saut, *Blowing up time-dependent solutions of the planar, Chern-Simons gauged nonlinear Schrödinger equation*, Nonlinearity **8** (1995), no. 2, 235–253. MR1328596

[2] L. Bergé, A. De Bouard, and J.-C. Saut, *Collapse of Chern-Simons-gauged matter fields*, Phys. Rev. Lett. **74** (1995), no. 20, 3907–3911, DOI 10.1103/PhysRevLett.74.3907. MR1329717

[3] J. Bourgain and W. Wang, *Construction of blowup solutions for the nonlinear Schrödinger equation with critical nonlinearity*, Ann. Scuola Norm. Sup. Pisa Cl. Sci. (4) **25** (1997), no. 1-2, 197–215 (1998). Dedicated to Ennio De Giorgi. MR1655515

[4] N. Burq, F. Planchon, J. G. Stalker, and A. S. Tahvildar-Zadeh, *Strichartz estimates for the wave and Schrödinger equations with the inverse-square potential*, J. Funct. Anal. **203** (2003), no. 2, 519–549, DOI 10.1016/S0022-1236(03)00238-6. MR2003358

[5] J. Byeon, H. Huh, and J. Seok, *Standing waves of nonlinear Schrödinger equations with the gauge field*, J. Funct. Anal. **263** (2012), no. 6, 1575–1608, DOI 10.1016/j.jfa.2012.05.024. MR2948224

[6] J. Byeon, H. Huh, and J. Seok, *On standing waves with a vortex point of order N for the nonlinear Chern-Simons-Schrödinger equations*, J. Differential Equations **261** (2016), no. 2, 1285–1316, DOI 10.1016/j.jde.2016.04.004. MR3494398

[7] S. S. Chern and J. Simons, *Characteristic forms and geometric invariants*, Ann. of Math. (2) **99** (1974), 48–69, DOI 10.2307/1971013. MR353327

[8] K. S. Chou and T. Y.-H. Wan, *Asymptotic radial symmetry for solutions of $\Delta u + e^u = 0$ in a punctured disc*, Pacific J. Math. **163** (1994), no. 2, 269–276. MR1262297

[9] B. Dodson, *Global well-posedness and scattering for the mass critical nonlinear Schrödinger equation with mass below the mass of the ground state*, Adv. Math. **285** (2015), 1589–1618, DOI 10.1016/j.aim.2015.04.030. MR3406535

[10] R. Donninger, M. Huang, J. Krieger, and W. Schlag, *Exotic blowup solutions for the u^5 focusing wave equation in $\mathbb{R}^3$*, Michigan Math. J. **63** (2014), no. 3, 451–501, DOI 10.1307/mmj/1409932630. MR3255688

[11] G. Dunne, *Self-dual chern-simons theories*, Lecture Notes in Physics Monographs, vol. 36, Springer-Verlag Berline Heidelberg, 1995.

[12] R. T. Glassey, *On the blowing up of solutions to the Cauchy problem for nonlinear Schrödinger equations*, J. Math. Phys. **18** (1977), no. 9, 1794–1797, DOI 10.1063/1.523491. MR460850

[13] S. Gustafson, K. Nakanishi, and T.-P. Tsai, *Asymptotic stability, concentration, and oscillation in harmonic map heat-flow, Landau-Lifshitz, and Schrödinger maps on $\mathbb{R}^2$*, Comm. Math. Phys. **300** (2010), no. 1, 205–242, DOI 10.1007/s00220-010-1116-6. MR2725187

[14] H. Huh, *Blow-up solutions of the Chern-Simons-Schrödinger equations*, Nonlinearity **22** (2009), no. 5, 967–974, DOI 10.1088/0951-7715/22/5/003. MR2501032

[15] H. Huh, *Energy solution to the Chern-Simons-Schrödinger equations*, Abstr. Appl. Anal., posted on 2013, Art. ID 590653, 7, DOI 10.1155/2013/590653. MR3035224

[16] H. Huh and J. Seok, *The equivalence of the Chern-Simons-Schrödinger equations and its self-dual system*, J. Math. Phys. **54** (2013), no. 2, 021502, 5, DOI 10.1063/1.4790487. MR3076362

[17] R. Jackiw and S.-Y. Pi, *Classical and quantal nonrelativistic Chern-Simons theory*, Phys. Rev. D (3) **42** (1990), no. 10, 3500–3513, DOI 10.1103/PhysRevD.42.3500. MR1084552

[18] R. Jackiw and S.-Y. Pi, *Soliton solutions to the gauged nonlinear Schrödinger equation on the plane*, Phys. Rev. Lett. **64** (1990), no. 25, 2969–2972, DOI 10.1103/PhysRevLett.64.2969. MR1056846

[19] R. Jackiw and S.-Y. Pi, *Time-dependent Chern-Simons solitons and their quantization*, Phys. Rev. D (3) **44** (1991), no. 8, 2524–2532, DOI 10.1103/PhysRevD.44.2524. MR1132645

[20] R. Jackiw and S.-Y. Pi, *Self-dual Chern-Simons solitons*, Progr. Theoret. Phys. Suppl. **107** (1992), 1–40, DOI 10.1143/PTPS.107.1. Low-dimensional field theories and condensed matter physics (Kyoto, 1991). MR1194691

[21] J. Jendrej, *Construction of two-bubble solutions for the energy-critical NLS*, Anal. PDE **10** (2017), no. 8, 1923–1959, DOI 10.2140/apde.2017.10.1923. MR3694010

[22] J. Jendrej, *Construction of type II blow-up solutions for the energy-critical wave equation in dimension 5*, J. Funct. Anal. **272** (2017), no. 3, 866–917, DOI 10.1016/j.jfa.2016.10.019. MR3579128

[23] J. Jendrej, *Nonexistence to two-bubbles with opposite signs for the radial energy-critical wave equation*, Ann. Sc. Norm. Super. Pisa Cl. Sci. (5) **18** (2018), no. 2, 735–778. MR3801295

[24] J. Jendrej, *Construction of two-bubble solutions for energy-critical wave equations*, Amer. J. Math. **141** (2019), no. 1, 55–118, DOI 10.1353/ajm.2019.0002. MR3904767

[25] J. Jendrej and A. Lawrie, *Two-bubble dynamics for threshold solutions to the wave maps equation*, Invent. Math. **213** (2018), no. 3, 1249–1325, DOI 10.1007/s00222-018-0804-2. MR3842064

[26] R. Killip, D. Li, M. Visan, and X. Zhang, *Characterization of minimal-mass blowup solutions to the focusing mass-critical NLS*, SIAM J. Math. Anal. **41** (2009), no. 1, 219–236, DOI 10.1137/080720358. MR2505858

[27] R. Killip, T. Tao, and M. Visan, *The cubic nonlinear Schrödinger equation in two dimensions with radial data*, J. Eur. Math. Soc. (JEMS) **11** (2009), no. 6, 1203–1258, DOI 10.4171/JEMS/180. MR2557134

[28] J. Krieger and W. Schlag, *Non-generic blow-up solutions for the critical focusing NLS in 1-D*, J. Eur. Math. Soc. (JEMS) **11** (2009), no. 1, 1–125, DOI 10.4171/JEMS/143. MR2471133

[29] J. Krieger, W. Schlag, and D. Tataru, *Renormalization and blow up for charge one equivariant critical wave maps*, Invent. Math. **171** (2008), no. 3, 543–615, DOI 10.1007/s00222-007-0089-3. MR2372807

[30] J. Krieger, E. Lenzmann, and P. Raphaël, *On stability of pseudo-conformal blowup for L^2-critical Hartree NLS*, Ann. Henri Poincaré **10** (2009), no. 6, 1159–1205, DOI 10.1007/s00023-009-0010-2. MR2557200

[31] J. Krieger and W. Schlag, *Full range of blow up exponents for the quintic wave equation in three dimensions*, J. Math. Pures Appl. (9) **101** (2014), no. 6, 873–900, DOI 10.1016/j.matpur.2013.10.008. MR3205646

[32] J. Krieger, W. Schlag, and D. Tataru, *Slow blow-up solutions for the $H^1(\mathbb{R}^3)$ critical focusing semilinear wave equation*, Duke Math. J. **147** (2009), no. 1, 1–53, DOI 10.1215/00127094-2009-005. MR2494455

[33] A. Lawrie, S.-J. Oh, and S. Shahshahani, *Self-dual Chern-Simons-Schrödinger equation*, unpublished, 1–9.

[34] D. Li and X. Zhang, *On the rigidity of solitary waves for the focusing mass-critical NLS in dimensions $d \geq 2$*, Sci. China Math. **55** (2012), no. 2, 385–434, DOI 10.1007/s11425-012-4359-1. MR2886544

[35] Z. M. Lim, *Large data well-posedness in the energy space of the Chern-Simons-Schrödinger system*, J. Differential Equations **264** (2018), no. 4, 2553–2597, DOI 10.1016/j.jde.2017.10.026. MR3737847

[36] B. Liu and P. Smith, *Global wellposedness of the equivariant Chern-Simons-Schrödinger equation*, Rev. Mat. Iberoam. **32** (2016), no. 3, 751–794, DOI 10.4171/RMI/898. MR3556051

[37] B. Liu, P. Smith, and D. Tataru, *Local wellposedness of Chern-Simons-Schrödinger*, Int. Math. Res. Not. IMRN **23** (2014), 6341–6398, DOI 10.1093/imrn/rnt161. MR3286341

[38] Y. Martel, *Asymptotic N-soliton-like solutions of the subcritical and critical generalized Korteweg-de Vries equations*, Amer. J. Math. **127** (2005), no. 5, 1103–1140. MR2170139

[39] Y. Martel, F. Merle, and P. Raphaël, *Blow up for the critical gKdV equation. II: Minimal mass dynamics*, J. Eur. Math. Soc. (JEMS) **17** (2015), no. 8, 1855–1925, DOI 10.4171/JEMS/547. MR3372073

[40] Y. Martel, F. Merle, and P. Raphaël, *Blow up for the critical gKdV equation III: exotic regimes*, Ann. Sc. Norm. Super. Pisa Cl. Sci. (5) **14** (2015), no. 2, 575–631. MR3410473

[41] Y. Martel and P. Raphaël, *Strongly interacting blow up bubbles for the mass critical nonlinear Schrödinger equation* (English, with English and French summaries), Ann. Sci. Éc. Norm. Supér. (4) **51** (2018), no. 3, 701–737, DOI 10.24033/asens.2364. MR3831035

[42] F. Merle, *Determination of blow-up solutions with minimal mass for nonlinear Schrödinger equations with critical power*, Duke Math. J. **69** (1993), no. 2, 427–454, DOI 10.1215/S0012-7094-93-06919-0. MR1203233

[43] F. Merle and P. Raphael, *Sharp upper bound on the blow-up rate for the critical nonlinear Schrödinger equation*, Geom. Funct. Anal. **13** (2003), no. 3, 591–642, DOI 10.1007/s00039-003-0424-9. MR1995801

[44] F. Merle, *Construction of solutions with exactly k blow-up points for the Schrödinger equation with critical nonlinearity*, Comm. Math. Phys. **129** (1990), no. 2, 223–240. MR1048692

[45] F. Merle and P. Raphael, *On universality of blow-up profile for L^2 critical nonlinear Schrödinger equation*, Invent. Math. **156** (2004), no. 3, 565–672, DOI 10.1007/s00222-003-0346-z. MR2061329

[46] F. Merle and P. Raphael, *The blow-up dynamic and upper bound on the blow-up rate for critical nonlinear Schrödinger equation*, Ann. of Math. (2) **161** (2005), no. 1, 157–222, DOI 10.4007/annals.2005.161.157. MR2150386

[47] F. Merle and P. Raphael, *Profiles and quantization of the blow up mass for critical nonlinear Schrödinger equation*, Comm. Math. Phys. **253** (2005), no. 3, 675–704, DOI 10.1007/s00220-004-1198-0. MR2116733

[48] F. Merle and P. Raphael, *On a sharp lower bound on the blow-up rate for the L^2 critical nonlinear Schrödinger equation*, J. Amer. Math. Soc. **19** (2006), no. 1, 37–90, DOI 10.1090/S0894-0347-05-00499-6. MR2169042

[49] F. Merle, P. Raphaël, and I. Rodnianski, *Blowup dynamics for smooth data equivariant solutions to the critical Schrödinger map problem*, Invent. Math. **193** (2013), no. 2, 249–365, DOI 10.1007/s00222-012-0427-y. MR3090180

[50] F. Merle, P. Raphaël, and J. Szeftel, *The instability of Bourgain-Wang solutions for the L^2 critical NLS*, Amer. J. Math. **135** (2013), no. 4, 967–1017, DOI 10.1353/ajm.2013.0033. MR3086066

[51] S.-J. Oh and F. Pusateri, *Decay and scattering for the Chern-Simons-Schrödinger equations*, Int. Math. Res. Not. IMRN **24** (2015), 13122–13147, DOI 10.1093/imrn/rnv093. MR3436140

[52] G. Perelman, *On the formation of singularities in solutions of the critical nonlinear Schrödinger equation*, Ann. Henri Poincaré **2** (2001), no. 4, 605–673, DOI 10.1007/PL00001048. MR1852922

[53] P. Raphaël, *Stability of the log-log bound for blow up solutions to the critical non linear Schrödinger equation*, Math. Ann. **331** (2005), no. 3, 577–609, DOI 10.1007/s00208-004-0596-0. MR2122541

[54] P. Raphaël and I. Rodnianski, *Stable blow up dynamics for the critical co-rotational wave maps and equivariant Yang-Mills problems*, Publ. Math. Inst. Hautes Études Sci. **115** (2012), 1–122, DOI 10.1007/s10240-011-0037-z. MR2929728

[55] P. Raphaël and R. Schweyer, *Stable blowup dynamics for the 1-corotational energy critical harmonic heat flow*, Comm. Pure Appl. Math. **66** (2013), no. 3, 414–480, DOI 10.1002/cpa.21435. MR3008229

[56] P. Raphaël and R. Schweyer, *Quantized slow blow-up dynamics for the corotational energy-critical harmonic heat flow*, Anal. PDE **7** (2014), no. 8, 1713–1805, DOI 10.2140/apde.2014.7.1713. MR3318739

[57] P. Raphaël and J. Szeftel, *Existence and uniqueness of minimal blow-up solutions to an inhomogeneous mass critical NLS*, J. Amer. Math. Soc. **24** (2011), no. 2, 471–546, DOI 10.1090/S0894-0347-2010-00688-1. MR2748399

[58] I. Rodnianski and J. Sterbenz, *On the formation of singularities in the critical* $\mathrm{O}(3)$ *σ-model*, Ann. of Math. (2) **172** (2010), no. 1, 187–242, DOI 10.4007/annals.2010.172.187. MR2680419

[59] A. Stefanov, *Strichartz estimates for the Schrödinger equation with radial data*, Proc. Amer. Math. Soc. **129** (2001), no. 5, 1395–1401, DOI 10.1090/S0002-9939-00-05722-1. MR1814165

[60] T. Tao, *Spherically averaged endpoint Strichartz estimates for the two-dimensional Schrödinger equation*, Comm. Partial Differential Equations **25** (2000), no. 7-8, 1471–1485, DOI 10.1080/03605300008821556. MR1765155

[61] M. I. Weinstein, *Nonlinear Schrödinger equations and sharp interpolation estimates*, Comm. Math. Phys. **87** (1982/83), no. 4, 567–576. MR691044

[62] M. I. Weinstein, *Modulational stability of ground states of nonlinear Schrödinger equations*, SIAM J. Math. Anal. **16** (1985), no. 3, 472–491, DOI 10.1137/0516034. MR783974

Editorial Information

To be published in the *Memoirs*, a paper must be correct, new, nontrivial, and significant. Further, it must be well written and of interest to a substantial number of mathematicians. Piecemeal results, such as an inconclusive step toward an unproved major theorem or a minor variation on a known result, are in general not acceptable for publication.

Papers appearing in *Memoirs* are generally at least 80 and not more than 200 published pages in length. Papers less than 80 or more than 200 published pages require the approval of the Managing Editor of the Transactions/Memoirs Editorial Board. Published pages are the same size as those generated in the style files provided for $\mathcal{A}_{\mathcal{M}}\mathcal{S}$-LaTeX.

Information on the backlog for this journal can be found on the AMS website starting from `http://www.ams.org/memo`.

A Consent to Publish is required before we can begin processing your paper. After a paper is accepted for publication, the Providence office will send a Consent to Publish and Copyright Agreement to all authors of the paper. By submitting a paper to the *Memoirs*, authors certify that the results have not been submitted to nor are they under consideration for publication by another journal, conference proceedings, or similar publication.

Information for Authors

Memoirs is an author-prepared publication. Once formatted for print and on-line publication, articles will be published as is with the addition of AMS-prepared frontmatter and backmatter. Articles are not copyedited; however, confirmation copy will be sent to the authors.

Initial submission. The AMS uses Centralized Manuscript Processing for initial submissions. Authors should submit a PDF file using the Initial Manuscript Submission form found at `www.ams.org/submission/memo`, or send one copy of the manuscript to the following address: Centralized Manuscript Processing, MEMOIRS OF THE AMS, 201 Charles Street, Providence, RI 02904-2294 USA. If a paper copy is being forwarded to the AMS, indicate that it is for *Memoirs* and include the name of the corresponding author, contact information such as email address or mailing address, and the name of an appropriate Editor to review the paper (see the list of Editors below).

The paper must contain a *descriptive title* and an *abstract* that summarizes the article in language suitable for workers in the general field (algebra, analysis, etc.). The *descriptive title* should be short, but informative; useless or vague phrases such as "some remarks about" or "concerning" should be avoided. The *abstract* should be at least one complete sentence, and at most 300 words. Included with the footnotes to the paper should be the 2020 *Mathematics Subject Classification* representing the primary and secondary subjects of the article. The classifications are accessible from `www.ams.org/msc/`. The Mathematics Subject Classification footnote may be followed by a list of *key words and phrases* describing the subject matter of the article and taken from it. Journal abbreviations used in bibliographies are listed in the latest *Mathematical Reviews* annual index. The series abbreviations are also accessible from `www.ams.org/msnhtml/serials.pdf`. To help in preparing and verifying references, the AMS offers MR Lookup, a Reference Tool for Linking, at `www.ams.org/mrlookup/`.

Electronically prepared manuscripts. The AMS encourages electronically prepared manuscripts, with a strong preference for $\mathcal{A}_{\mathcal{M}}\mathcal{S}$-LaTeX. To this end, the Society has prepared $\mathcal{A}_{\mathcal{M}}\mathcal{S}$-LaTeX author packages for each AMS publication. Author packages include instructions for preparing electronic manuscripts, samples, and a style file that generates the particular design specifications of that publication series.

Authors may retrieve an author package for *Memoirs of the AMS* from `www.ams.org/journals/memo/memoauthorpac.html`. The *AMS Author Handbook* is available in PDF format from the author package link. The author package can also be obtained free of charge by sending email to `tech-support@ams.org` or from the Publication Division,

American Mathematical Society, 201 Charles St., Providence, RI 02904-2294, USA. When requesting an author package, please specify the publication in which your paper will appear. Please be sure to include your complete mailing address.

After acceptance. The source files for the final version of the electronic manuscript should be sent to the Providence office immediately after the paper has been accepted for publication. The author should also submit a PDF of the final version of the paper to the editor, who will forward a copy to the Providence office.

Accepted electronically prepared files can be submitted via the web at `www.ams.org/submit-book-journal/`, sent via FTP, or sent on CD to the Electronic Prepress Department, American Mathematical Society, 201 Charles Street, Providence, RI 02904-2294 USA. TEX source files and graphic files can be transferred over the Internet by FTP to the Internet node `ftp.ams.org` (130.44.1.100). When sending a manuscript electronically via CD, please be sure to include a message indicating that the paper is for the *Memoirs*.

Electronic graphics. Comprehensive instructions on preparing graphics are available at `www.ams.org/authors/journals.html`. A few of the major requirements are given here.

Submit files for graphics as EPS (Encapsulated PostScript) files. This includes graphics originated via a graphics application as well as scanned photographs or other computer-generated images. If this is not possible, TIFF files are acceptable as long as they can be opened in Adobe Photoshop or Illustrator.

Authors using graphics packages for the creation of electronic art should also avoid the use of any lines thinner than 0.5 points in width. Many graphics packages allow the user to specify a "hairline" for a very thin line. Hairlines often look acceptable when proofed on a typical laser printer. However, when produced on a high-resolution laser imagesetter, hairlines become nearly invisible and will be lost entirely in the final printing process.

Screens should be set to values between 15% and 85%. Screens which fall outside of this range are too light or too dark to print correctly. Variations of screens within a graphic should be no less than 10%.

Any graphics created in color will be rendered in grayscale for the printed version unless color printing is authorized by the Managing Editor and the Publisher. In general, color graphics will appear in color in the online version.

Inquiries. Any inquiries concerning a paper that has been accepted for publication should be sent to `memo-query@ams.org` or directly to the Electronic Prepress Department, American Mathematical Society, 201 Charles St., Providence, RI 02904-2294 USA.

Editors

SELECTED PUBLISHED TITLES IN THIS SERIES

1404 **Luchezar Stoyanov,** Spectral Properties of Ruelle Transfer Operators for Regular Gibbs Measures and Decay of Correlations for Contact Anosov Flows, 2023

1403 **Akinari Hoshi, Ming-chang Kang, and Aiichi Yamasaki,** Multiplicative Invariant Fields of Dimension ≤ 6, 2023

1402 **Roberto Frigerio and Marco Moraschini,** Gromov's Theory of Multicomplexes with Applications to Bounded Cohomology and Simplicial Volume, 2023

1401 **Jang Soo Kim, Kyu-Hwan Lee, and Se-jin Oh,** Weight Multiplicities and Young Tableaux Through Affine Crystals, 2023

1400 **Tarek M. Elgindi and In-Jee Jeong,** On Singular Vortex Patches, I: Well-Posedness Issues, 2023

1399 **Fan Gao, Freydoon Shahidi, and Dani Szpruch,** Local Coefficients and Gamma Factors for Principal Series of Covering Groups, 2023

1398 **David A. Croydon, Tsuyoshi Kato, Makiko Sasada, and Satoshi Tsujimoto,** Dynamics of the Box-Ball System with Random Initial Conditions via Pitman's Transformation, 2023

1397 **Anand P. Patel and Ashvin A. Swaminathan,** Inflectionary Invariants for Isolated Complete Intersection Curve Singularities, 2023

1396 **Margaret Bilu,** Motivic Euler Products and Motivic Height Zeta Functions, 2023

1395 **Sebastian Casalaina-Martin, Samuel Grushevsky, Klaus Hulek, and Radu Laza,** Cohomology of the Moduli Space of Cubic Threefolds and Its Smooth Models, 2023

1394 **Valentin Blomer, Étienne Fouvry, Emmanuel Kowalski, Philippe Michel, Djordje Milićević, and Will Sawin,** The Second Moment Theory of Families of L-Functions–The Case of Twisted Hecke L-Functions, 2023

1393 **Óscar Domínguez and Sergey Tikhonov,** Function Spaces of Logarithmic Smoothness: Embeddings and Characterizations, 2023

1392 **André Gil Henriques, David Penneys, and James Tener,** Planar Algebras in Braided Tensor Categories, 2023

1391 **Tiago J. Fonseca,** Higher Ramanujan Equations and Periods of Abelian Varieties, 2023

1390 **Tomasz Downarowicz and Guohua Zhang,** Symbolic Extensions of Amenable Group Actions and the Comparison Property, 2023

1389 **João C. A. Barata, Christian D. Jäkel, and Jens Mund,** The $\mathcal{P}(\varphi)_2$ Model on de Sitter Space, 2023

1388 **Guozhen Lu, Jiawei Shen, and Lu Zhang,** Multi-Parameter Hardy Spaces Theory and Endpoint Estimates for Multi-Parameter Singular Integrals, 2023

1387 **François Ledrappier and Lin Shu,** The Regularity of the Linear Drift in Negatively Curved Spaces, 2023

1386 **Zhaobing Fan, Chun-Ju Lai, Yiqiang Li, Li Luo, and Weiqiang Wang,** Affine Hecke Algebras and Quantum Symmetric Pairs, 2023

1385 **D. Kaledin,** Mackey Profunctors, 2022

1384 **Matt Bainbridge, John Smillie, and Barak Weiss,** Horocycle Dynamics: New Invariants and Eigenform Loci in the Stratum $\mathcal{H}(1,1)$, 2022

1383 **Chris Kottke and Michael Singer,** Partial Compactification of Monopoles and Metric Asymptotics, 2022

1382 **Jenny Fuselier, Ling Long, Ravi Ramakrishna, Holly Swisher, and Fang-Ting Tu,** Hypergeometric Functions Over Finite Fields, 2022

1381 **Henri Berestycki and Grégoire Nadin,** Asymptotic Spreading for General Heterogeneous Fisher-KPP Type Equations, 2022

For a complete list of titles in this series, visit the
AMS Bookstore at **www.ams.org/bookstore/memoseries/**.